CAMBRIDGE MONOGRAPHS IN
EXPERIMENTAL BIOLOGY
No. 14

EDITORS:
T. A. BENNET-CLARK
P. W. BRIAN, G. M. HUGHES
GEORGE SALT (*General Editor*)
V. B. WIGGLESWORTH

THE NEUROCHEMISTRY OF ARTHROPODS

THE SERIES

THE
NEUROCHEMISTRY OF
ARTHROPODS

J . E . TREHERNE, Ph.D.

A.R.C. Unit of Insect Physiology,
Department of Zoology, University of Cambridge

CAMBRIDGE
AT THE UNIVERSITY PRESS
1966

PUBLISHED BY
THE SYNDICS OF THE CAMBRIDGE UNIVERSITY PRESS

Bentley House, 200 Euston Road, London, N.W. 1
American Branch: 32 East 57th Street, New York, N.Y. 10022
West African Office: P.M.B. 5181, Ibadan, Nigeria

©

CAMBRIDGE UNIVERSITY PRESS

1966

Printed in Great Britain at the University Printing House, Cambridge
(Brooke Crutchley, University Printer)

LIBRARY OF CONGRESS CATALOGUE
CARD NUMBER: 66-12306

260281

CONTENTS

PREFACE

PROGRESS in neurophysiological research has been largely achieved through the use of a limited number of experimental preparations. Such preparations as cephalopod giant axons, crustacean stretch receptors, frog sciatic nerve-muscle and mammalian brain slices have yielded valuable insight into various aspects of the functioning of nerve cells. These results are usually considered in relation to the mammalian nervous system, and there have been few attempts to integrate them with the particular structure and organization of the nervous systems of invertebrate groups. This trend is in many ways unfortunate, for many physiological processes can be more readily understood when considered in the context of the relatively simple invertebrate nervous systems. The underlying principles of the so-called 'blood-brain barrier' phenomena can, for example, be more clearly defined in the arthropod central nervous system than in the more complex mammalian brain. Recent investigations have also shown that the nervous system of arthropod species may be particularly suited to enable analyses of behavioural acts to be carried out at the neurophysiological level. It is of some importance, therefore, to arrive at an understanding of the basic chemical and physical events occurring within the nervous system of this group of animals. Finally it should be mentioned that the very urgent problem of the mechanism of insecticide resistance must ultimately be stated in neurochemical terms and necessarily involves an understanding of the normal events occurring within the nervous tissues.

In this monograph an account is given of the chemistry of the nervous system of the largest of the animal groups. In particular an attempt has been made to consider the chemical and physiological events in relation to the specialized structure of the arthropod nervous tissues, and the first chapter is a brief outline of the organization of the arthropod nervous system. The subsequent chapters deal with the chemistry of axonal and synaptic transmission, with energy expenditure and with various aspects of the metabolic activities of arthropod nervous tissues.

Unfortunately limitations of space have prevented any discussion of neurosecretory phenomena, neither has it been possible to devote any pages to the toxicological aspects of neurochemistry. In most chapters an attempt has been made to give a fairly complete coverage of the existing literature, although in some, notably chapters 1 and 10, the limited number of pages available has necessitated some selection of the published work.

I am particularly grateful to several of my friends and colleagues for their kindness in reading and commenting on various portions of the manuscript of this monograph. I have been especially helped in this way by the kind efforts of Dr E. H. Colhoun, Professor E. Florey, Dr T. I. Shaw, Dr D. S. Smith and Mr F. P. W. Winteringham. I should also like to acknowledge my indebtedness to Sir Vincent Wigglesworth, whose unfailing interest and support has enabled me to become acquainted with the subject of invertebrate neurochemistry. Finally, I should like to thank Mrs K. J. Leonard for typing the manuscript and Mr R. T. Hughes for checking the references in this volume.

I am indebted to the authors who have allowed me to reproduce their diagrams and data and also to the following publishing companies for their permission to reproduce illustrations: Academic Press Inc., The Company of Biologists Ltd, J. and A. Churchill Ltd, Pergamon Press, The Physiological Society, The Rockefeller Institute, John Wiley and Sons Inc., and the Wistar Institute of Anatomy and Biology.

Cambridge   J. E. T.
1965

CHAPTER I

The Organization of the Arthropod Nervous System

An adequate understanding of the metabolic processes occurring within the nervous system requires a knowledge of the structure and organization of the various nervous tissues. This introductory chapter is an outline of the structural organization of the arthropod nervous system, which will be integrated in succeeding chapters with what is known of the physiological processes of these tissues.

The primitive structure of the arthropod nervous system is generally held to be that of a dorsal brain with two longitudinal ventral nerve cords, containing a pair of ganglia for each somite, together with associated peripheral nerves. Such an arrangement is exhibited, for example, in Branchiopod crustaceans. In other arthropods varying degrees of fusion of the ganglia have occurred which are frequently associated with the union of various somites. In species such as the crayfish or in some insects it is, however, still possible to recognize the ladder-like arrangement of at least some portion of the ventral nerve cord. Extreme examples of the fusion of ganglia are found among Decapod crustaceans, some arachnids and such highly evolved insects as cyclorraphous dipterans.

These evolutionary developments, together with the great differences in size between species, have resulted in enormous variations in the dimensions of the central nervous structures. This variation is especially apparent in the ganglionic masses, which may vary from a fraction of a millimetre in diameter in various species of mites to dimensions which are to be measured in centimetres in arthropods such as Decapod crustaceans.

Although some crustacean groups, as exemplified by the Malacostraca, possess capillaries which penetrate into the nervous tissues, the ganglia of many arthropods are solid masses of tissue which, unlike vertebrate nervous structures, possess no internal channels enabling a circulation of body fluids to take place. In insects the tracheae and tracheoles penetrate into the central and peripheral nervous systems, carrying oxygen directly to the tissues. It is not known what proportion of the oxygen is carried in this way in those arachnids which possess tracheae; there are, however, no equivalent pathways in the nervous systems of other arthropods, so that in these species respiratory exchanges can take place only through capillaries, where these are developed, or through the surface of nervous structures.

The nerve sheath

The superficial connective tissue sheath (plate 1), through which in many species all the necessary exchanges of ions and molecules with the blood must take place, is a characteristic feature of the arthropod nervous system. Although the nerve sheath was described in insects by Michels and in a crustacean by Huxley in 1880, most of our knowledge is derived from some relatively recent investigations in insects. The connective tissue layer of the sheath is presumably equivalent to the vertebrate perineurium; there has, however, been some confusion in the terminology which has been used to describe the connective tissue and underlying cellular layer in arthropods. The synonomy which has been used in the description of the cellular and non-cellular portions of the sheath in insects is summarized by Ashhurst (1959) and Smith and Treherne (1963). In this monograph the outer, non-cellular portion of the sheath will be referred to as the 'neural lamella'; in the absence of any clarification as to the homologies of this part of the nervous system, the term 'perineurium' will be retained to describe the underlying cellular layer of the sheath.

Electron-micrographs of the neural lamella in *Periplaneta americana* (Hess, 1958; Smith and Treherne, 1963), *Rhodnius prolixus* (Smith and Wigglesworth, 1959) and *Locusta migratoria* (Ashhurst and Chapman, 1961) have revealed a complex structure consisting of a narrow outer homogeneous or finely granular region surrounding the remainder of the layer, which is filled with periodically banded fibrils. The presence of these fibrils confirms the results of X-ray diffraction studies (Rudall, 1955; Richards and Schneider, 1958), birefringence data (Baccetti, 1956, 1957) and histochemical investigations (Ashhurst, 1959, 1961a), all of which indicate the presence of collagen in this

2

layer of the nerve sheath. The latter histochemical studies suggest that the non-fibrous matrix of *Periplaneta* and *Locusta* probably consists of a neutral mucopolysaccharide, while in *Galleria* it is formed of an acid mucopolysaccharide (Ashhurst and Richards, 1964). The collagenous material is deployed in a multidirectional meshwork, with the long axes of the fibrils generally arranged tangentially with respect to the ganglion as a whole. Such an arrangement of collagen fibrils within the neural lamella makes this structure relatively inextensible. The collagen of the insect neural lamella resembles that from the rat-tail (Gray, 1960) except for the insertion of an extra doubled subperiod band between the single bands defining the limits of the macroperiod (Smith and Treherne, 1963).

The structure of the underlying cellular layer, or perineurium, has been described in light and electron-microscope studies on *Periplaneta americana* (Scharrer, 1939; Hess, 1958; Wigglesworth, 1960; Ashhurst, 1961a; Pipa, 1961; Smith and Treherne, 1963) and on *Rhodnius* (Wigglesworth, 1958). In the terminal abdominal ganglion of *Periplaneta* this layer has the form of a narrow epithelium laterally and is hypertrophied near the origin of the connectives and the insertion of the cercal nerves (Wigglesworth, 1960). The cells of the insect perineurium do not have a conspicuous basement membrane, while the outer plasma membrane is overlaid by the basal region of the neural lamella (plate 1). This layer contains numerous mitochondria which often have an aggregated distribution within the cell; they are frequently found in large numbers in the region immediately adjoining the surface of the neural lamella and in clusters deeper in the cytoplasm.

In the few other arthropods which have been investigated there appears to be no layer of specialized cells lying immediately beneath the neural lamella. In the isopod *Armadillidium vulgare* and the chilopod *Lithobius*, for example, the neurones are separated from the connective tissue layer only by thin cytoplasmic prolongations which originate in the neurone satellite cells (Trujillo-Cenóz, 1962).

Glial cells

In both a crustacean (*Armadillidium*) and a myriapod (*Lithobius*) (Trujillo-Cenóz, 1962), and in some insects (Wigglesworth, 1960; Pipa, 1961; Smith and Treherne, 1963), it is possible to

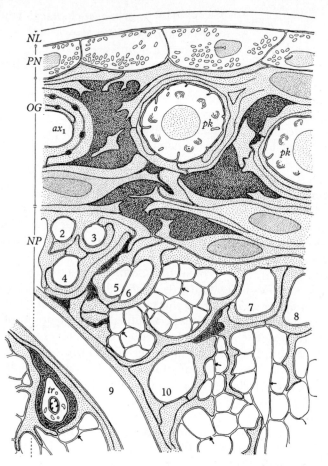

Fig. 1. Diagrammatic representation of the structure of the terminal abdominal ganglion of the cockroach *Periplaneta americana,* based on electron-microscope studies. The glial cytoplasm is indicated by light stippling and the extensive extracellular spaces by dark stippling. The ganglion is covered by the fibrous sheath, the neural lamella (NL), beneath which is situated a cellular layer, the perineurium (PN). The outer portion of the ganglion (OG), between the perineurium and the neuropile (NP), contains the neurone cell bodies or perikarya (pk) encapsulated by glial cell processes. A nerve fibre (ax_1) is shown surrounded by a concentric glial sheath. The inner glial cells of this region of the ganglion send processes into the neuropile. The extensive extracellular spaces situated towards the periphery of the ganglion represent the 'glial lacunar system' of Wigglesworth (1960).

The neuropile, which is the region of synaptic contact of the ganglion, contains both nervous and glial elements. Many of the axon profiles within the neuropile are ensheathed by glial cell processes (e.g. axons 2–10). Elsewhere there is a close apposition of axon plasma membranes in the absence of glial cell prolongations.

4

recognize two rather ill-defined regions in the remaining glial system: the peripheral layer, which contains cells that invest the neurone cell bodies, and the inner region, adjacent to the

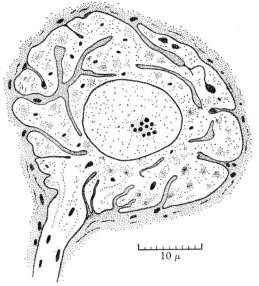

Fig. 2. Motor ganglion cells from the central nervous system of the insect *Rhodnius prolixus*, showing the multiple invaginations of the plasma membrane. The surrounding glial cytoplasm contains numerous mitochondria. (From Wigglesworth, 1959.)

surface of the neuropile, in which the cells send a complex system of processes into the deeper layers of the ganglion. In *Periplaneta* (fig. 1) these two regions of the glial system are partially separated by an irregular system of extracellular spaces, 'the glial lacunar system' (Wigglesworth, 1960). Details of the

---

In the latter cases the axon surfaces are separated by extracellular gaps in the region of 100–150 Å (arrows). In addition to this network of narrow channels between the axons spaces of similar dimensions occur between the axons and glial processes. Larger spaces are occasionally found in these deeper regions of the ganglion, either surrounding the tracheoles (*tr*) or lying between the glial processes (e.g. as between axons 4 and 5). The extracellular system of the neuropile is confluent with that of the more peripheral regions of the ganglion. (For purposes of clarity the region between the perineurium and the neuropile has been reduced, while the dimensions of the extracellular spaces between the axon branches and the glial processes in the neuropile have been exaggerated.) (From Smith and Treherne, 1963.)

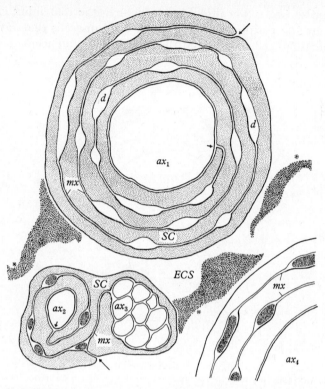

Fig. 3. Diagrammatic representation of the relationship between the axons and glial cell sheaths in the central nervous system of the cockroach *Periplaneta americana*. The larger axons (as at ax_1) are invested with a concentric Schwann cell sheath (SC), within which a mesaxon invagination (mx) extends from the surface (long arrow) and terminates as a sheath separated from the plasma membrane of the axon by a gap of c. 100–150 Å. The membranes of the mesaxon folds diverge at the point indicated by the short arrow. The spiral channels formed by the mesaxon folds are approximately 100–150 Å in width and contain occasional dilatations (d). These dilatations usually contain a homogeneous extracellular material. The smaller axons may be associated with separate mesaxon invaginations (e.g. as between arrows at ax_2) or a group of axons may be surrounded by a common Schwann cell (e.g. as at ax_3). (From Smith and Treherne, 1963.)

cytological organization of the glial cells of *Periplaneta* are given by Hess (1958) and Smith and Treherne (1963).

The intimate relation between the gliocyte processes and the surface of the neurone cell body was first demonstrated in the description of the 'trophospongium' by Holmgren (1900). These deep and irregular invaginations, which contain the narrow processes of the glial envelope, have since been found

6

in the crayfish *Cambarus* (Ross, 1915), the cladoceran *Leptodora* (Scharrer, 1964) and in the insects *Rhodnius* (Wigglesworth, 1959) (fig. 2) and *Periplaneta* (Hess, 1958; Smith and Treherne, 1963). The latter electron-microscope studies have shown that the apposed cell surfaces are produced into small-scale interdigitations.

There are varying degrees of association between the glial cell sheaths and their associated axons in arthropods. Species such as the scorpion *Bothriurus bonariensis* (Trujillo-Cenóz, 1962) or *Periplaneta americana* (Hess, 1958; Smith and Treherne, 1963) exhibit a typical tunicated condition in which the mesaxon folds form only a limited number of spirals (fig. 3). Some prawns and a few other crustaceans, on the other hand, possess nerve fibres which are almost as heavily myelinated as vertebrate axons of similar dimensions (Holmes, 1942). In the central axons illustrated in fig. 3 the mesaxon folds define an extracellular channel, of between 100 and 150 Å in width. This channel is interspersed with dilatations or lacunae which are generally filled with an electron-dense homogeneous material. In the larger peripheral nerves, such as the cercal nerve of the cockroach (Smith and Treherne, 1963), there are many axons of varying size and the axon sheaths are closely applied to each other. In the outer regions of the ganglia, however, the sheathed axons may be separated by fairly extensive extracellular spaces. In the smaller peripheral nerves one or more axons may be present. The Schwann cell or lemnoblast in these may be invaginated to form mesaxons at a number of points and may branch and reform along their course (Edwards, Ruska and de Harven, 1958*a*).

The neurone

Most of the recent literature on the neurone cell bodies and axons has been concerned with their distribution and pathways within the central and peripheral nervous systems of arthropods (cf. Wiersma, 1961; Bullock and Horridge, 1965). Some details of the cytology and ultrastructure of arthropod neurones are available for a scorpion, an isopod and a chilopod (Trujillo-Cenóz, 1962) and for various species of insects (Hess, 1958; Gray, 1960; Trujillo-Cenóz, 1959; Wigglesworth, 1960; Ashhurst, 1961*c*; Smith and Treherne, 1963).

The cell bodies of the motor or internunciary neurones lie at the periphery of the ganglion. The axon processes lead into the

7

central neuropile which is the site of synaptic transmission and integration in the arthropod central nervous system. The synapses which occur at the surface of the vertebrate cell bodies are precluded, in the cockroach at least, by the glial insulation of this portion of the neurone (Smith and Treherne, 1963).

The cytoplasm of the cell body of *Periplaneta* contains a great variety of inclusions (Hess, 1958; Smith and Treherne, 1963). Prominent amongst these were the highly ordered membrane associations, commonly known as dictyosomes or Golgi bodies, together with numerous mitochondria and neurofibrillae similar to those described in vertebrate perikarya (Palay and Palade, 1955). The cytoplasmic organization of the cell body changes abruptly at the axon cone (the region of transition between the perikaryon and the axon). In particular the axoplasm lacks the extensive system of granular cisternae and unattached ribosomes. A similar restriction of these elements to the perikaryon has also been noted in other arthropods by Trujillo-Cenóz (1962) and Gray (1960).

Studies on the physical properties of lobster giant axons indicate the presence of structural elements in the axoplasm which restrain the movements of oil droplets (Feldherr, 1958). Lobster axoplasm, however, appears to be much less rigid than that of squid giant axons, and its viscosity has been shown, by two different methods, to be in the region of 5·5 centipoises (Riesner, 1949; Tobias and Bryant, 1955). In lobster axons there is a vigorous Brownian movement, together with a vibratile movement of the filamentous structures scattered through the axoplasm (Feldherr, 1958).

In addition to mitochondria and various vesicular structures the axoplasm of several arthropods has been found to contain filamentous structures. Light-microscope studies have revealed the presence of a fibrillar system in both the axoplasm and perikaryon of insect neurones (Cajal and Sanchez, 1915; Monti, 1913; Beams and King, 1932; Pipa and Cook, 1958) which appear to correspond to vertebrate neurofilaments (cf. Palay and Palade, 1955). These structures have also been noted in electron-microscope studies of arthropod neurones (de Robertis and Schmitt, 1948; Trujillo-Cenóz, 1959, 1962; Smith and Treherne, 1963). The axoplasm in ganglia of *Periplaneta*, for example, contains narrow oriented tubules which are approximately 200 Å in diameter (Smith and Treherne, 1963).

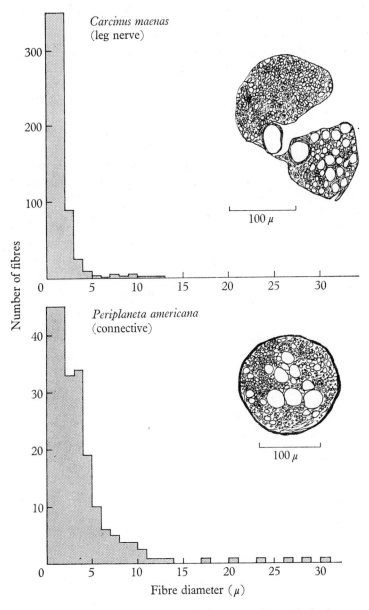

Carcinus maenas
(leg nerve)

100 μ

Periplaneta americana
(connective)

100 μ

Number of fibres

Fibre diameter (μ)

Fig. 4. The distribution of fibre size in the leg nerve of the crab *Carcinus maenas*, and an abdominal connective of the cockroach *Periplaneta americana*. The section of the *Carcinus* nerve is of a single bundle of fibres. (Drawn from the data and photomicrographs of Keynes and Lewis, 1951, and Roeder, 1953.)

9

Similar, but more sparsely distributed, structures are also found in the perikaryon in this species. Phase-contrast preparations have shown that the axoplasm of both large and small axons in *Periplaneta* contains unstained canals or fibrils, which are approximately 0·5 μ in diameter (Wigglesworth, 1960). These structures most probably correspond to bundles of neurofilaments. These filaments or tubules found in the neurone provide a possible structural pathway linking the metabolism of the perikaryon with that of the axon.

There is considerable variation in the diameter of the axons in arthropod nervous structures. In general, however, smaller fibres appear to be most numerous (fig. 4). The smaller fibres are largely the sensory ones, the motor fibres being fewer in number and larger in diameter (Roeder, 1953). In *Carcinus* peripheral nerve the area of the axon surfaces amounts to about 7500 cm²/g of tissue (Keynes and Lewis, 1951), while in *Maia* the membrane surface has been calculated to be as high as 10,000 cm²/g of tissue (Abbott, Hill and Howarth, 1958).

Synapses

It is evident from the fragmentary information available for arthropods that several of the cytological features of the regions of synaptic transmission in the neuropile are essentially similar to those in vertebrates (Palay, 1958; de Robertis, 1958, 1959). Now the neuropile of the arthropod central nervous system consists of a very complex association of axon branches and glial processes, in which the interpolation of glial cytoplasm frequently precludes the establishment of synaptic contact between adjacent axon processes. In the scorpion *Bothriurus bonariensis*, the isopod *Armadillidium vulgare* (Trujillo-Cenóz, 1962) and the crayfish (Robertson, 1964) as well as some species of insects (Hess, 1958; Trujillo-Cenóz, 1959, 1962; Smith and Treherne, 1963), however, the neuroglial processes may be absent in restricted regions so as to form a gap in which the closely applied axon surfaces are only separated by a distance of about 100 Å. Such regions represent possible sites for synaptic pathways in the central nervous system and it is significant that in many of these synaptic foci resembling those described by Palay (1958) have been observed in *Periplaneta* (Smith and Treherne, 1963). In these regions the axoplasm of the presumed presynaptic member contains clusters of vesicles (plate 2). These clusters are approximately 150–500 mμ in length and closely adjoin the axon membrane.

The fine structure of the inhibitory synapse at the crustacean stretch receptor neurone is essentially similar to that of excitatory synaptic endings (Peterson and Pepe, 1961). As with other synaptic regions the crayfish inhibitory synapse contains vesicular structures in the presynaptic terminal and extensive accumulations of mitochondria. The only unusual feature of this type of synapse is the presence of a variable latticework of 230 Å tubules in the connective tissue immediately adjacent to the inhibitory endings.

In the crayfish nerve cord the median-to-motor giant synapse is formed by the penetration of processes from the presynaptic giant fibre through the sheath of the post-synaptic fibre (Johnson, 1924; Robertson, 1953, 1961). Comparison with physiological studies on this preparation (Furshpan and Potter, 1959a; Wiersma, 1947) show that, unlike the vertebrate synapses which have been studied, it is the post-synaptic, rather than the presynaptic, axoplasm which contains the most striking collection of mitochondria and vesicular structures. It remains to be discovered whether this feature is confined to this peculiar form of synapse or whether it can be extended to those in the neuropile of other arthropods. This crayfish synapse is also unusual in that the usual 100–150 Å gap between unit membranes found in the axon-Schwann complex is completely absent in the synaptic membrane complex (Robertson, 1964).

The neuromuscular junction

Arthropods differ from vertebrates in that only a limited number of nerve fibres may innervate each muscle (Wiersma, 1941; Hoyle, 1957). In many instances a single arthropod muscle fibre can receive functional endings from two or three different motor axons (van Harreveld, 1939; Hoyle, 1955). More detailed accounts of the innervation of muscles in arthropods are given by Hoyle (1957) and Wiersma (1961).

Electron-micrographs show that the insect neuromuscular junction corresponds to that of the vertebrate motor end-plate in the close apposition of the plasma membranes of the axon branch and muscle fibre (Edwards et al., 1958a, b; Edwards, 1959, 1960; Smith, 1960, 1961). With the exception of the flight muscle of *Tenebrio* and *Aeschna* the basement membrane of the glial sheath joins that of the sarcolemma before the junction is reached, the terminal glial cell layer being retained as the superficial covering over the axon terminal. In these two species the lemnoblast is absent, the naked axon either lying in a groove, as in *Aeschna*, or being fully invaginated into one fibre surface, as in *Tenebrio*. The terminal axoplasm of the insect neuromuscular junction contains vesicles and mitochondria much as in the vertebrate motor end-plate (plate 2). There is, however, much variability in the cytology of the post-synaptic region in insects and it is only the organization of the terminal axoplasm and the close apposition of the synaptic surfaces which are essentially similar to the vertebrate condition (Smith and Treherne, 1963).

CHAPTER 2

Water

THERE is considerable variation in the water content of nervous tissues in arthropods (table 1). From the evidence available it seems that there is a much greater proportion of solids in the tissues of the insect central nervous system than in the crustaceans or the arachnid which have been studied. This is particularly evident in the nerve cord of the stick insect, *Carausius morosus*, in which solid materials account for nearly 40 % of the fresh weight of the nerve cord. There is no simple relation between the water content of the tissues and the osmotic concentration of the blood. The stick insect, for example, which has the most dilute blood has also the lowest water content in the nervous tissues of any of the species listed in table 1. This lack of correlation might be expected in view of the presence of the inextensible nerve sheath, which might tend to prevent the cells from coming into osmotic equilibrium with the blood. There is, however, an appreciable osmotic uptake of water in hypotonic solutions by peripheral nerve of the spider crab, *Libinia canaliculata* (Guttman, 1939), although it is not clear to what extent this is restrained by the peripheral nerve sheath. Such a departure from a simple relation between tissue water content and the osmotic pressure of the blood is also likely to result from variations in the molecular weight of the constituents of the tissues from the various species of arthropods.

The contribution of the various constituent electrolytes to the osmotic balance in the peripheral nerve of *Carcinus* has been estimated by Lewis (1952) (fig. 5). The greatest osmotic activity was exhibited by the monovalent inorganic cations, the next most conspicuous group of electrolytes being the amino acids, especially aspartate and glutamate. It seems therefore that these amino acids, besides balancing essential internal cations, are also necessary additional solutes which contribute towards the osmotic equilibrium of the nerve fibres.

TABLE 1. *The water content of some arthropod nervous tissues compared with the osmotic concentration of the blood*

Species	Preparation	Blood $\Delta°$ C	Water content of nervous tissues (% fresh wt.)
CRUSTACEA			
Homarus americanus	Peripheral nerve	1·811–1·880 (Cole, 1940)	89·2 ± 3·5 (Nevis, 1958)
'Lobster'	Ventral nerve cord	—	84·6 (Brante, 1949)
	Claw nerve	—	86·1
Maia squinado	Peripheral nerve	2·04 (Margaria, 1931)	86·8 (Cowan, 1934)
Cancer pagurus	Peripheral nerve	—	84·8 (Cowan, 1934)
Carcinus maenas	Peripheral nerve	1·29–1·82 (Schlieper, 1929)	86·5 (Lewis, 1952)
ARACHNIDA			
Limulus polyphemus	Leg nerve	1·01–1·880 (Cole, 1940)	82·8 (Young, 1938)
INSECTA			
Periplaneta americana	Abdominal nerve cord	0·897 (Treherne, 1961 a)	75·5 ± 0·9 (Tobias, 1948 b) 73·3 ± 0·9 (Treherne, unpublished)
	Terminal ganglion	—	77·8 ± 1·8 (Treherne, 1962 b)
Romalea microptera	Nerve cord	—	78·9 ± 2·4 (Tobias, 1948 b)
Carausius morosus	Nerve cord	0·53 (Ramsay, 1955)	60·6 ± 1·8 (Treherne, 1965 b)

The exchange of tritiated water between the haemolymph and the nervous tissues has been studied in the peripheral nerve of the lobster *Homarus americanus* (Nevis, 1958), and the nerve cord of the cockroach *Periplaneta americana* (Treherne, 1962 a, b). In these investigations it was assumed that the behaviour of the tritiated water molecules was similar to that of unlabelled ones. This assumption appears to be justified as far as permeability studies are concerned, for the free self-diffusion constants for water labelled with deuterium, tritium or ^{18}O were similar over the whole temperature range tested (Wang, Robinson and Edelmann, 1953).

In the above investigations the fluxes were studied by measuring the escape of labelled water from preparations which had been made radioactive either by injection of the radioisotope into the blood or by soaking *in vitro*. In both cases the efflux was found to occur as a two-stage process; an initial rapid escape eventually giving way to a second slower exponential phase (fig. 6). In *Homarus* the rapidly exchanging fraction accounted for 35·2 % and in *Periplaneta* 21·6 % of the tissue water. These rapidly exchanging molecules were identified as those associated with the extracellular spaces in these tissues. In the case of

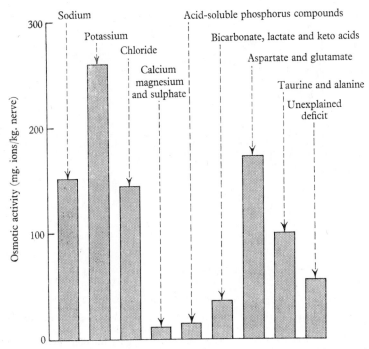

Fig. 5. The osmotic activity of the principal solutes in leg nerves of the crab *Carcinus maenas*. (From the data of Lewis, 1952.)

Periplaneta this identification receives some corroboration from the fact that the volume of the rapidly exchanging water was similar to that of the measured inulin space (of 18·2 %) and from the observation that the equivalent rapidly exchanging sodium fraction was qualitatively different from the slowly exchanging (cellular) one in that it was unaffected by the presence of metabolic inhibitors (p. 24) (Treherne, 1961 e, 1962 b).

The apparent volume of 21·6 % for the extracellular water in the tissues of the central nervous system of *Periplaneta*, together with the figure of 15·2 % for the nerve cord of *Carausius morosus* (Treherne, 1965 *b*), might be considered relatively high for these tissues, for most electron-micrographs of the vertebrate central nervous system have revealed only very restricted extracellular spaces consisting of the gaps of a few hundred Å between adjacent plasma membranes and accounting for only a few per

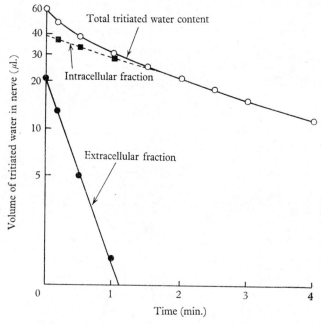

Fig. 6. The efflux of tritiated water from isolated peripheral nerve of the lobster *Homarus americanus* when washed in normal physiological solution after equilibration in a radioactive solution. The rapidly exchanging fraction (extracellular fraction) was obtained by extrapolation of the main curve (open circles) to zero time. (After Nevis, 1958.)

cent of the tissue water (cf. Horstmann and Meves, 1959). In the preceding chapter, however, evidence was discussed which showed that the nerve cord of *Periplaneta americana* contains much larger extracellular spaces, especially the extensive glial lacunar system (fig. 1) (Smith and Treherne, 1963). A recent description has also been given of electron-micrographs which exhibit more extensive extracellular spaces even in vertebrate cerebral tissues (van Harreveld and Crowell, 1964). The comparative aspects of the distribution of extracellular water in the

16

tissues of the central nervous system have been reviewed by Treherne (1962d, 1965c).

The influx of water into the cells of lobster peripheral nerve has been shown to be related to the osmotic concentration of the external medium (fig. 7) (Nevis, 1958). These exchanges are, however, not affected by the presence of metabolic inhibitors, while the changes in efflux rates at various temperatures show that these exhibit the characteristics of passive diffusion processes with an apparent activation energy of 2·5 kcal. It is

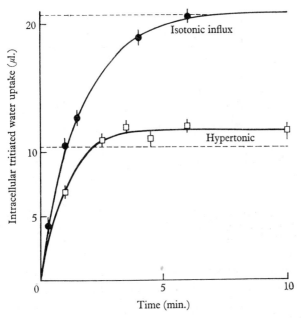

Fig. 7. Influx of tritiated water into leg nerves of the lobster *Homarus americanus* from isotonic and hypertonic balanced salt solutions. (After Nevis, 1958.)

apparent, therefore, that any active transport of water into or out of the lobster nerve fibre must be very small in comparison to the exchanges taking place by diffusion processes. Calculation of the filtration/permeability ratio ($P_f/P_d = 20 \pm 7$ at 14° C), according to the theories of Koefoed-Johnsen and Ussing (1953) and Pappenheimer, Renkin and Borrero (1951), indicates that these exchanges of water molecules were taking place through water-filled spaces in the axon membrane with 'effective pore radii' of approximately 16 ± 4 Å. Villegas and Villegas (1960), however, have criticized the mathematical treatment adopted in Nevis's analysis and point out that a pore radius as large as 16 Å would be difficult to reconcile with the low sodium and potassium conductance of the axolemma. In their work on the squid axon Villegas and Villegas postulate a maximum pore size of 8·5 Å.

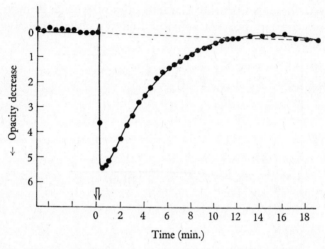

Fig. 8. The change in opacity of the leg nerve of the crab *Maia squinado* following stimulation at 50/sec for 15 sec. The period of stimulation is indicated by the twin arrows. (After Hill, 1950.)

The exchanges of water molecules between fibres and the surrounding medium can be related to the electrical activity of the nerve. In both *Carcinus maenas* (Hill and Keynes, 1949) and *Maia squinado* (Hill, 1950) the peripheral nerve fibres undergo a reversible change in opacity on stimulation (fig. 8). The rapid decrease in opacity appears to be due to an increase in diameter of the fibres: dilution of the external medium, for example, causes a swelling of the fibres which is accompanied by equivalent changes in opacity. The inward movement of water following stimulation probably results from increases in osmotic concentration of the axoplasm caused by the ionic changes taking place in active nerve. The exchange of external sodium for the internal potassium ions on stimulation would, for example, be expected to produce an increase in osmotic pressure, for the entry of sodium is also accompanied by an influx of chloride ions with a consequent increase in the osmotic pressure of the fibres. These observations also accord with some earlier work on the effects of the alteration of the external sodium/potassium ratio, which have been shown to produce net inward movements of water in the axons of the spider crab (Shanes, 1946). The small initial increase in opacity noted in *Carcinus* nerve fibres is difficult to account for. It is conceivable, however, that the initial influx of sodium might occur without the associated hydration water so as to cause a transient volume increase (Hill, 1950).

CHAPTER 3

Inorganic Ions

THE electrical activity of the nervous system is profoundly influenced by the concentrations of inorganic ions within the nerve cells and in the fluid immediately surrounding them. In this chapter an attempt has been made to describe the distribution of inorganic ions within various arthropod nervous tissues and, in particular, to define the nature and composition of the extracellular environment of the nerve cells. An account is also given of the exchanges taking place between the body fluids and the nervous tissues, and between the neurones and the extracellular fluid at rest and during electrical activity.

The composition of the blood

The exchanges and distribution of inorganic ions in tissues of the nervous system are of especial interest in the Arthropoda, both for the profound variations in the composition of the blood which occur between different groups within the phylum and for the large fluctuations in the ionic concentration which have been shown to take place in the body fluids of individual animals. In all of the Crustacea-Malacostraca which have been investigated sodium chloride accounts for the greater part of the osmotic concentration of the blood, potassium, calcium, magnesium and organic substances forming only a small proportion of the whole (cf. Florkin, 1960; Robertson, 1960; Lockwood, 1962). The body fluids of an arachnid (*Tegenaria atrica*), a chilopod (*Lithobius*) (Croghan, 1959) and an apterygote insect (*Petrobius maritimus*) (Lockwood and Croghan, 1959) are essentially similar in these respects to those of crustaceans. The body fluids of all these arthropods approximate to vertebrate blood in the possession of high concentrations of sodium and the relatively low levels of potassium and divalent ions. In the pterygote insects, however, there is great variation in the ionic composition of the blood and the sodium/potassium ratio tends to be consistently lower than in the other groups within the phylum (cf. Prosser and Brown, 1961; Wyatt, 1961; Shaw and Stobbart, 1963; Sutcliffe, 1963). In certain orders, such as Coleoptera, Lepidoptera, Hymenoptera and Phasmida, the absolute level of sodium in the blood may be exceedingly low. In the beetle *Timarchia tenebroisa*, for example, the sodium concentration has been recorded as 1·6 mM/l; with levels of potassium at 46·9 mM/l, calcium at 36·1 mM/l and magnesium at 79 mM/l (Duchâteau, Florkin and Leclercq, 1953). The high concentration of potassium ions in the blood of the lepidopteran *Telea polyphemus* does not appear to be correlated with any significant binding of the ion, although between 15 and 20% of the calcium and magnesium ions were associated with macromolecules (Carrington and Tenney, 1959).

Although many arthropods regulate the ionic concentration of the blood, some species show fairly dramatic fluctuations with changing external conditions. In the brine shrimp, *Artemia salina*, for example, the sodium level of the blood was shown to vary between approximately 75 and 300 mM/l with different dilutions of the external medium (Croghan, 1958). The ionic content of the blood may also depend upon the nutritional state of the animal; in the locust, for example, the potassium level of fed insects was twice that of starved individuals (Hoyle, 1954). Changes in the potassium level of the blood resulting from modifications in the diet can also be related to the activity and behaviour of adult cockroaches (Pichon and Boistel, 1963).

The inorganic ion content of arthropod nervous tissues

Table 2 summarizes the available information on the ionic content of arthropod nervous tissues. In all the species the gross internal potassium concentration exceeds that of sodium. In the three insects which have been investigated the absolute level of sodium ions is lower than in the two crustaceans for which figures are available; this difference can be correlated with the lower level of this ion in the blood of the insect species. The stick insect, *Carausius morosus*, is remarkable for the fact that the tissue sodium exceeds that of the blood (in terms of tissue water the actual ratio is 5·09—Treherne, 1965d). The nerve cord potassium of this insect is also higher than that of any of the other arthropods, being over three times that of *Romalea* despite the similar level of this ion in the blood of the two species.

21

TABLE 2. *The gross contents of inorganic ions in nervous tissues and haemolymph of some arthropods* (mM/kg tissue)

Species	Tissue	Na	K	Mg	Ca	Cl	PO_4	HCO_3	SO_4	Author
CRUSTACEA										
Homarus americanus	Peripheral nerve	—	203·0	—	65·0*	107·0	3·7	—	—	Fenn et al. 1934 and Engel and Gerard, 1935
	Haemolymph	231·9	8·0	5·4	13·9	270·0	—	—	4·0	Cole, 1940
Libinia marginata	Peripheral nerve	163·0	158·0	—	67·5*	134·0	—	—	—	Fenn et al. 1934
	Haemolymph	—	—	—	—	—	—	—	—	
Carcinus maenas	Peripheral nerve	152·0	260·0	11·5	6·5	145·0	15·0	9·0	9·0	Keynes and Lewis, 1951 and Lewis, 1952
	Haemolymph	531·0	12·3	19·5	13·3	557·0	—	—	16·4	Webb, 1940
Cancer pagurus	Peripheral nerve	—	134·4	—	—	—	—	—	—	Cowan, 1934
	Haemolymph	502·2	11·9	27·1	13·8	508·9	—	—	24·0	Robertson, 1939
Maia squinado	Peripheral nerve	—	132·2	—	—	—	—	—	—	Cowan, 1934
	Haemolymph	488·0	12·4	44·1	13·6	554·0	—	—	16·5	Robertson, 1953
ARACHNIDA										
Limulus polyphemus	Peripheral nerve	—	123·4	—	—	—	—	—	—	Young, 1938
	Haemolymph	303·1	8·7	24·6	6·3	308·0	—	—	5·6	Cole, 1940
INSECTA										
Periplaneta americana	Nerve cord	60·9	107·4	2·9	—	—	—	—	—	Tobias, 1948a
	Nerve cord	75·6	132·1	—	—	—	—	—	—	Treherne, 1961a
	Haemolymph	156·5	7·7	5·3	4·2	144·4	—	—	—	Asperen and Esch, 1956
Romalea microptera	Nerve cord	69·5	89·0	—	—	—	—	—	—	Tobias, 1948b
	Haemolymph	56·5	17·9	—	—	—	—	—	—	Tobias, 1948b
Carausius morosus	Nerve cord	63·8	313·4	22·1	30·2	—	—	—	—	Treherne, 1965b
	Haemolymph	15·0	18·0	53·0	7·5	101·0	16·0	5·0	—	Wood, 1957

* These high values should be compared with those of 6·7 and 6·3 mM/kg for crab and lobster nerve (Tipton, 1934).

The concentrations summarized in table 2 include both extracellular and intracellular ion fractions. There has been no direct measurement of the intracellular concentrations by determinations on extruded axoplasm, such as has been carried out with squid axons (e.g. Steinback and Spiegelma, 1943). An estimate of the intracellular electrolyte concentration has been made for *Carcinus* nerve, using a measured polysaccharide space of 0.242 ± 0.011 litre/kg (Lewis, 1952). The estimated concentrations within the cells were markedly different from those of whole nerve; the values for intracellular sodium being lower and those for potassium being higher than in the whole nerve (table 3). These values were calculated on the assumption that the extracellular concentrations of the electrolytes were the same as in the external medium. Experiments with various radioisotopes on insect nerve cords (Treherne, 1961 *e*, 1962 *a, b*, 1965 *b*) have shown that, as in mammalian peripheral nerve (Krnjević, 1955), the concentrations of the extracellular cations were higher and the anions lower than in the outside solution. The true intracellular values may thus deviate from those summarized in table 3. Estimates of the intracellular concentrations of various cations have been made from consideration of the kinetics of their exchange with the external medium. These will be described in the remainder of the chapter.

TABLE 3. *The probable intracellular concentration of electrolytes in peripheral nerve of* Carcinus maenas. *These values are based on a mean measured polysaccharide space of* 0.242 ± 0.011 *and an estimated maximum space of* 0.272 *litre/kg tissue* (*Lewis, 1952*)

	Concentration		Quantity in 1·0 kg nerve		Estimated intracellular concentration
Electrolyte	In nerve (mM/kg)	In Ringer (mM/l)	Extracellular (mM)	Intracellular (mM)	(mM/kg water)
Na	152	495	119–134	33–18	53–30
K	260	11	3	257	412–432
Cl	145	532	128–144	17–1	27
Aspartate and glutamate	173	—	—	173	276–291
Alanine and taurine	98	—	—	98	157–165

The movements of sodium between the blood and the tissues of the nerve cord has been studied in the insects *Periplaneta americana* and *Carausius morosus*. As will be seen from table 2 the latter species has a relatively low level of sodium in the blood, this ion being an order more concentrated in the blood of the cockroach.

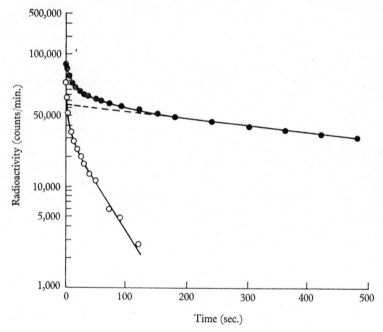

Fig. 9. The loss of [24]Na from a whole ligatured abdominal nerve cord (closed circles) of the cockroach *Periplaneta americana*, which had been made radioactive by soaking for 20 min in a physiological solution containing the radioisotope. The fast component of the main curve (open circles) was obtained by subtraction from the straight line extrapolated to zero time. (From Treherne, 1962 *b*.)

Despite these differences in concentration in the blood the uptake of sodium into the tissues of the nerve cord was found to occur at approximately similar rates in the two species (Treherne, 1961 *a*, 1965 *b*).

The escape of labelled sodium ions into non-radioactive external medium was found to occur as a two-stage process in the nerve cords of *Periplaneta* (Treherne, 1961 *e*, 1962 *b*) and *Carausius* (Treherne, 1965 *b*) (fig. 9). In both species the escape of sodium ions in the slowly exchanging fraction was found to be reduced in the presence of low concentrations of cyanide or 2:4-dinitrophenol and to be unaffected by the removal of substantial portions of the fibrous and cellular nerve sheath in *Periplaneta* terminal abdominal ganglion (Treherne, 1961 *d*). The efflux of sodium in the rapidly exchanging fraction, on the other hand, was not affected by the presence of metabolic inhibitors. On the basis of this evidence and the previous data for the efflux of tritiated water and inulin, discussed in the previous chapter, it was postulated as a first approximation that the rapidly exchanging fraction represented sodium contained in the extracellular fluid and that the slowly exchanging one consisted of the intracellular ions. The form of the rapid efflux was essentially similar to that predicted by the equations of Hill (1928) for the diffusion of molecules from the extracellular space in a mass of muscle, in which the initially complex efflux eventually follows a simple exponential with a half-time given by:

$$t_{0.5} = 0.118 r_0^2 / D',$$

where r_0 is the radius of the tissue and D' the diffusion constant in the extracellular space. The above equation can be modified for a complex structure such as an arthropod nerve cord by representing it thus:

$$t_{0.5} = a1/D',$$

where a is a constant. The term $1/t_{0.5}$ is thus proportional to the diffusion constant in the extracellular spaces of this complex structure. The data obtained for *Periplaneta* (Treherne, 1962 *b*) showed that there was not a simple relation between the movements of sodium ions and other substances and their free diffusion constants. The broken line in fig. 10 shows the relationship which would be expected if the diffusion in the extracellular fluid was unrestricted. Clearly the sodium and other cations diffused more slowly than would be expected from their free diffusion constants. The non-electrolytes, chloride and tritiated water, can, however, be related by a line such as the continuous one shown in fig. 10. This relationship shows the existence of considerable restriction to the free diffusion of the sodium and other cations in the extracellular spaces of the nerve cord of *Periplaneta*. Such a restriction to the movement of cations could result from encounters with fixed anion groups (p. 27) within

25

the extracellular system or from the presence of peripheral cationic groups.

The use of the inulin space or the rapidly exchanging tritiated water fraction as a measure of the extracellular water showed that the concentration of the rapidly exchanging sodium ions exceeded that of the outside solution in both *Periplaneta* and

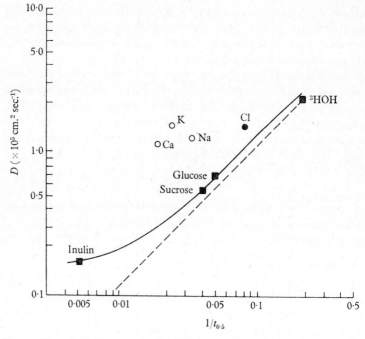

Fig. 10. The relation between the free diffusion constant (D) and the value $1/t_{0.5}$ (which is proportional to the diffusion constant in the extracellular spaces) obtained for various test substances in their diffusion from the extracellular system of the cockroach abdominal nerve cord. (From Treherne, 1962*b*.)

Carausius nerve cord (tables 4 and 5). Comparison of the ion ratios between the external medium and the extracellular fluid in *Periplaneta* indicated that, with the possible exception of potassium, the elevated cation and low chloride levels in the extracellular fluid were due to a Donnan equilibrium with the blood (table 4). The substances containing the free anion groups which cause this Donnan effect in the extracellular spaces of the central nervous system have not been conclusively identified. In experiments in which ganglia were desheathed before being soaked in a solution containing ^{22}Na it was shown that the concentrations of the ions fell to a lower level than would be expected on the basis of fixed anion groups alone. It seems likely, therefore, that some of the cations in the extracellular fluid may be associated with large molecules in solution which are dispersed on desheathing the ganglion (Treherne, 1962 b). The possibility also exists that the anion groups in the extracellular spaces of the cockroach central nervous system (p. 4) may be those associated with collagen or the extracellular acid mucopolysaccharide demonstrated in ganglia of this insect by Ashhurst (1961 b).

TABLE 4. *The distribution of sodium, potassium, calcium and chloride ions between the external medium and the extracellular and intracellular components of the nerve cord of* Periplaneta. *These estimated concentrations are based on a volume of extracellular water of 21·6 %, calculated from the rapidly exchanging tritiated water fraction in this preparation. (Data from Treherne, 1961 a and 1962 b)*

Ion	External concentration (mM/l)	Estimated extracellular (mM/l)	Estimated intracellular (mM/l)	Ion ratio	
Na	157·0	283·6	67·2	$\dfrac{Na_{out}}{Na_{extracellular}}$	$= 0·55$
K	12·3	17·1	225·1	$\dfrac{K_{out}}{K_{extracellular}}$	$= 0·71$
Ca	4·5	17·6	14·7	$\left(\dfrac{Ca_{out}}{Ca_{extracellular}}\right)^{\frac{1}{2}}$	$= 0·51$
Cl	184·0	106·7	—	$\dfrac{Cl_{extracellular}}{Cl_{out}}$	$= 0·58$

The extracellular sodium level in the nerve cord of *Carausius* exceeded the concentration in the blood by a much greater factor than in the cockroach, the estimated extracellular/haemolymph ratio for this ion being 10·6 (table 5). The distribution of this ion between the external medium and the extracellular fluid was also difficult to relate to a simple Donnan equilibrium such as has been shown to exist in *Periplaneta*. In *in vitro* experiments or in poisoned preparations the level of the rapidly exchanging sodium fraction fell dramatically as compared with the intact nerve cord (table 5). The high extracellular concentration of sodium was, in fact, found to depend upon an active uptake of this ion into the nerve cord (Treherne, 1965 b). In the presence of low concentrations of metabolic inhibitors the initial extremely rapid influx of this ion was reduced as compared with normal preparations (fig. 11). The extremely rapid influx of sodium ions into the extracellular fluid can be very

TABLE 5. *The calculated concentrations of the rapidly exchanging ion fractions in the nerve cord of* Carausius. *The estimates of the extracellular concentration are based on the measured inulin space of the nerve cord* (*Treherne, 1965 b*)

Ion	Type of experiment	Concentration in haemolymph or external solution (mM/l)	Calculated extracellular concentration (mM/l)	Ion ratio	
Na	*in vivo*	20·1	212·4*		= 10·6
	in vitro	15·0	78·7†	$\dfrac{Na_{extracellular}}{Na_{out}}$	= 5·2
	in vitro (poisoned)	15·0	18·5†		= 1·2
K	*in vitro*	18·0	67·1†	$\dfrac{K_{extracellular}}{K_{out}}$	= 3·7
	in vitro (poisoned)	18·0	72·5†		= 4·0
Ca	*in vitro*	7·5	14·4†	$\left(\dfrac{Ca_{extracellular}}{Ca_{out}}\right)^{\frac{1}{2}}$	= 1·4
	in vitro (poisoned)	7·5	15·0†		= 1·4
Mg	*in vitro*	50·0	94·4†	$\left(\dfrac{Mg_{extracellular}}{Mg_{out}}\right)^{\frac{1}{2}}$	1·4
Cl	*in vitro*	133·0	80·7–214·9†		

* Based on *in vivo* inulin space of 91·8 ml/kg tissue.
† Based on *in vitro* inulin space of 146·1 ml/kg tissue.

approximately described by a permeability constant (K_{in}) of $9·86 \times 10^{-5}$ cm^{-2} sec^{-1}. This constant is no greater than that of $7·60 \times 10^{-4}$ cm^{-2} sec^{-1} calculated for the anal papillae of mosquito larvae (Treherne, 1965 b) from the data of Treherne (1954) and Stobbart (1959). It is clear from these results that this active uptake of sodium maintains a relatively high concentration of these ions in the fluid immediately surrounding the nerve cells, and thus reproduces by this physiological mechanism an environment which approximates to the body fluids of other

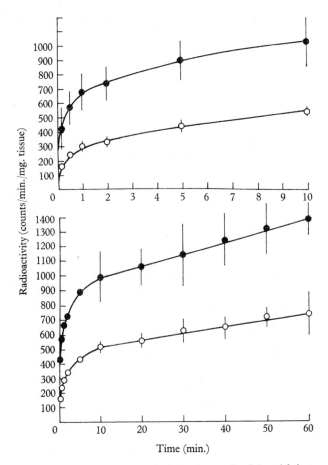

Fig. 11. The uptake of ^{22}Na by isolated nerve cords of the stick insect, *Carausius morosus*, soaked in radioactive physiological solution of similar concentration to the blood (p. 22). The closed circles represent the uptake in normal solution; the open circles that measured in a solution containing 0·5 mM/l 2:4-dinitrophenol. The vertical lines represent the extent of twice the standard error. The upper figure illustrates the initial uptake of ^{22}Na; the lower one shows the uptake obtained in the complete experiment. (From Treherne, 1965b.)

unspecialized animals. The estimated extracellular and intra-cellular concentrations of sodium in the nerve cord of this insect are summarized in table 6.

The extracellular sodium level in the poisoned preparations still exceeded that of the external medium by a factor of 1·2 (table 5). The ion ratios shown in table 5 indicate that this remaining extracellular sodium may be a result of a Donnan equilibrium with the external medium, for the value for the poisoned preparation approximates to those for calcium and magnesium ions.

TABLE 6. *The estimated cellular and extracellular ion concentrations in the nerve cord of* Carausius. *The values for sodium are based on* in vivo *studies, those for the remaining cations involve the extrapolation of* in vitro *results to the tissue concentrations shown in table 2.* (*Treherne, 1965 b.*)

Cation	Concentration in haemolymph (mM/l)	Estimated concentration in extracellular fluid (mM/l)	Estimated intracellular concentration (mM/l)
Na	20·1	212·4	86·3
K	33·7	124·5	555·8
Ca	6·4	12·2	61·8
Mg	61·8	117·4	10·7

The low intracellular concentration of sodium relative to that in the extracellular fluid in the nerve cords of *Periplaneta* and *Carausius* (tables 4 and 6) appears to be maintained, as in squid axon (Hodgkin and Keynes, 1955), by an active extrusion of these ions from the cells. In the presence of low concentrations of cyanide and 2:4-dinitrophenol, the slow exponential efflux of sodium ions from the nerve cord of *Periplaneta* (Treherne, 1961 e, 1962 b) and of *Carausius* (Treherne, 1965 b) was significantly reduced (fig. 12 a). The action of dinitrophenol on these insect axons suggests that, as with squid axons in which sodium extrusion has also been shown to depend upon the presence of ATP (Caldwell *et al.*, 1960), the sodium pump is dependent upon the action of oxidative phosphorylation (see p. 70). The escape of sodium from the nerve cord of *Periplaneta* was unaffected by the replacement of the external sodium by xylose or choline chloride, but was significantly reduced in the absence of the relatively low concentration of potassium ions (Treherne, 1961 b, c) (fig. 12 b). Such a coupling of sodium and potassium movements has been demonstrated in several cells and tissues (cf. Hodgkin, 1958), and has led to the hypothesis that it might be a result of a mechanism by which

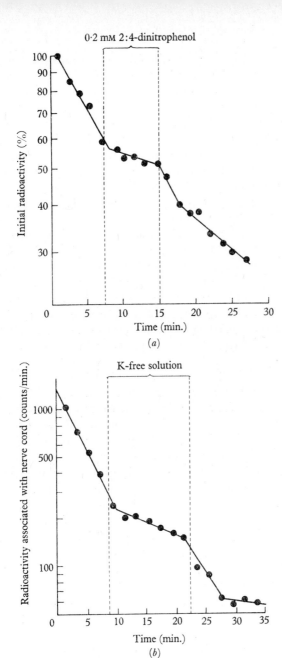

Fig. 12. The effect of dilute dinitrophenol (a) and potassium-free bathing solution (b) on the escape of ^{24}Na from the isolated abdominal nerve cord of the cockroach *Periplaneta americana*. This sodium efflux represents the slow (intracellular) component illustrated in fig. 9.

one sodium ion is extruded for each potassium ion absorbed (Harris, 1954; Hodgkin and Keynes, 1954). In the experiments with *Periplaneta* nerve cord the rate of sodium efflux in the potassium-free solution did not fall to the same extent as that in the presence of the metabolic inhibitors. In this case, as with *Sepia* axons (Hodgkin, 1958), it is very difficult to be sure that the concentration of potassium immediately surrounding the nerve cells had not been raised by the leakage of potassium which might allow some limited coupled exchanges of sodium and potassium to continue. It is, therefore, not possible to postulate whether the coupling between sodium efflux and potassium influx is a rigid or a partial one in this arthropod.

There have been no measurements of sodium fluxes during activity (either by using the isotope method or by measurement of the changes of the total tissue content of this ion following prolonged stimulation) for the axons of any arthropods. Indirect estimates for the entry of sodium into arthropod giant axons can be made, using the values for the membrane electrical constants (table 9), as will be discussed in the next chapter (p. 52).

Exchanges of potassium ions

Despite earlier claims of an incomplete exchange of ^{42}K in some nerve and muscle (Hevesy and Hahn, 1941; Rothenberg, 1950) no evidence of any appreciable amounts of 'bound' potassium was found in experiments on the peripheral nerve of *Carcinus* (Keynes and Lewis, 1951) or in insect nerve cords (Treherne, 1961 a; 1965 b). In all of these experiments with arthropod nervous tissues the exchange of potassium ions approached 100 %.

The escape of ^{42}K from the nerve cord of *Periplaneta americana* (Treherne, 1962 b) and *Carausius morosus* (Treherne, 1965 b) was found to occur as a two-stage process, an initial rapid escape eventually giving way to a low exponential leakage. As with sodium, the diffusion of potassium ions through the extracellular system in the cockroach nerve cord was restricted as compared with movements of chloride ions, water or non-electrolytes (Treherne, 1962 b) (fig. 10).

The concentration of potassium in the rapidly exchanging extracellular fraction in *Periplaneta* nerve cord exceeded the level of the ion in the external medium. This effect can, as a first approximation, be attributed to the Donnan equilibrium with the outside solution (table 4). In *Carausius*, however, the rapidly exchanging potassium fraction clearly could not be related to a simple Donnan equilibrium with the external medium when compared with the level of sodium ions in poisoned preparations and the concentrations of calcium and magnesium in the extracellular fluid (table 5). The level of the rapidly exchanging potassium also appeared to be independent of the concentration of ^{42}K in the slowly exchanging (intracellular) fraction and no metabolic processes seemed to be involved in its uptake into the extracellular fluid (Treherne, 1965 *b*). This effect is difficult to explain, but it is conceivable that it could result from some specific anion groups which might be expected to produce a reduction in the mean activity coefficient of the potassium ions in the extracellular fluid.

The slow escape of intracellular potassium ions from whole *Carcinus* nerve did not follow a simple exponential decline with time and the apparent exchange constant for the loss of ^{42}K in inactive solution tended to increase with time (fig. 13) (Keynes and Lewis, 1951). It is suggested that such an effect could result from the non-uniformity in exchange constants of the individual fibres, resulting either from the wide range of fibre size or from variations in membrane conductance. A similar effect might also result from swelling of the individual fibres in this isolated preparation. The average value for the exchange constant for this preparation was 0.49 hr^{-1}, which corresponds to a half-time of 1.4 hr. The results of experiments with bundles containing about six isolated 30μ axons yielded a rate constant for ^{42}K loss of 0.37 hr^{-1}, from which it was possible to estimate a potassium efflux of 22×10^{-12} Mcm^{-2} sec^{-1} and an influx of 19×10^{-12} Mcm^{-2} sec^{-1}. The efflux of potassium from the nerve cord of *Periplaneta* took place with a half-time of 0.257 hr (Treherne, 1962 *b*), which corresponds to an exchange constant of 2.69 hr^{-1}. This efflux from the insect nerve cord can be compared with the first portion of the graph illustrated in fig. 13 for whole *Carcinus* nerve, in which the exchange constant was 0.62 hr^{-1}. As the distribution of fibre size is very approximately similar in the nerve cord of *Periplaneta* and the peripheral nerve of *Carausius* (fig. 4) it seems evident that the exchanges of potassium ions take place somewhat more rapidly between the nerve cells and the extracellular space in the insect preparation. The escape of ^{42}K from *Carcinus* nerve can also be related to

33

external potassium concentration of the bathing solution (fig. 14), an effect which parallels the changes in membrane conductance at different potassium concentrations (Hodgkin, 1947).

Evidence has already been discussed which shows that the slow exponential phase of sodium efflux in *Periplaneta* is related to the potassium concentration of the bathing medium and extracellular fluid, probably as a result of some sort of 'linked'

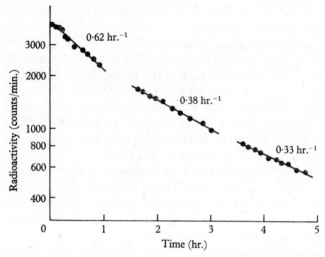

Fig. 13. A semi-logarithmic plot of the leakage of ^{42}K from a whole leg nerve of the crab *Carcinus maenas* on washing in non-radioactive Ringer solution. The figures against each set of counts are the calculated exchange constants for this portion of the curve. (After Keynes and Lewis, 1951.)

ion pump in which the active sodium extrusion is associated with a corresponding uptake of potassium. There is also some evidence of an active uptake of potassium in the nerve cord of *Carausius* (Treherne, 1965b). In these experiments the initial rapid uptake of ^{42}K was found not to be affected by the presence of metabolic inhibitors, although the slower absorption of the ion by the nerve cord was reduced by the addition of poisons to the bathing solution. As the initial ionic exchanges have been shown to be those taking place with the extracellular fluid it seems most likely that the effect of the metabolic inhibitors on the slower uptake of potassium by the tissues was to reduce the intracellular absorption of the ion. The demonstration of an

34

increased net leakage of potassium from isolated leg nerves of the crab *Libinia emarginata* under anoxic conditions also points to a relation between the metabolism and the movements of these ions in arthropod nerve (Shanes, 1950). The action of glucose in retarding potassium loss during oxygen lack can be related to its effect in reducing anoxic depolarization in crab nerve (Cowan, 1934; Shanes, 1949).

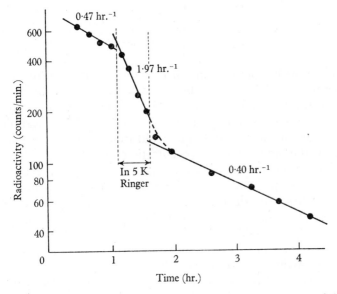

Fig. 14. The effect of external potassium concentration on the escape of ^{42}K from whole leg nerve of the crab *Carcinus maenas* when washed in inactive solution. The nerves were first washed in normal Ringer solution and then exposed to a solution with five times the normal potassium concentration before being returned to the normal solution. The figures for each portion of the graph represent the calculated exchange constants for ^{42}K efflux. (After Keynes and Lewis, 1951.)

The experiments on the kinetics of potassium exchange (Keynes and Lewis, 1951; Treherne, 1962*b*, 1965*b*), together with estimates made from the measured inulin space (Lewis, 1952), indicate that the ion is present in relatively high concentration in the intracellular fraction in arthropod nervous tissues. The estimated intracellular concentrations of potassium are given in tables 3, 4 and 6. The very low intracellular chloride concentration in crab nerve suggests that, as in other excitable tissues, the high potassium content of the cells is due, in part at least, to a Donnan equilibrium with the extracellular fluid. Calculation of the ion ratios from the data given in table 3 yields the following values:

$$\frac{Na_{in}}{Na_{out}} = \frac{53}{495} = 0 \cdot 107,$$

$$\frac{K_{in}}{K_{out}} = \frac{412}{11} = 37 \cdot 4,$$

$$\frac{Cl_{out}}{Cl_{in}} = \frac{532}{27} = 19 \cdot 7.$$

By analogy with other cells and from data on other arthropods (Treherne, 1961 b, c, d, e, 1962 b, 1965 b), it is reasonable to attribute the low values for the sodium concentrations to the action of the sodium pump. The high value for the potassium ratio relative to that for chloride could result from a low estimate for the intracellular concentration of the anion. The possibility also exists that the presence of a linked ion pump, which secretes potassium into the nerve cells, might cause a departure from the distribution which would be expected in the event of a simple Donnan equilibrium between the intracellular and extracellular fluid.

The exchanges of potassium ions associated with the electrical activity of arthropod nerve were first investigated over thirty years ago when Cowan (1934) demonstrated that *Maia* peripheral nerve lost an appreciable amount of its internal potassium on stimulation. These results were also confirmed for *Limulus* leg nerve (Young, 1938). The extra potassium lost from *Maia* nerve during stimulation was calculated to be about $2 \cdot 92 \times 10^{-10}$ M/g/impulse. Using the estimated value of 10,000 cm^2 membrane surface per gram of nerve (p. 10), this becomes $2 \cdot 92 \times 10^{-14}$ Mcm^{-2} impulse^{-1}. This is very much lower than the value of $7 \cdot 6 \times 10^{-12}$ Mcm^{-2} impulse^{-1} obtained by measuring the escape of ^{42}K in stimulated nerve with this species (Abbott, 1958) and with the figure of $2 \cdot 4 \times 10^{-12}$ Mcm^{-2} impulse^{-1} obtained with *Carcinus maenas* (fig. 15) (Keynes, 1951). The value for the escape of potassium from stimulated *Carcinus* nerve arrived at by indirect methods was also found to exceed the earlier estimates, averaging approximately $1 \cdot 7 \times 10^{-12}$ M cm^{-2} impulse^{-1} (Hodgkin and Huxley, 1947). These differences obtained using different methods could result from the fact that in the experiments of Cowan (1934) the fibres were stimulated to fatigue and may have become inexcitable before the end of the experiment.

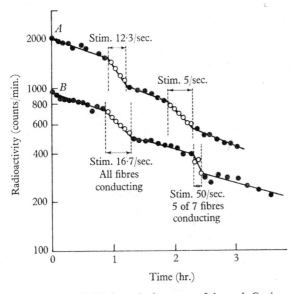

Fig. 15. The escape of ^{42}K from the leg nerve of the crab *Carcinus maenas* during stimulation in normal non-radioactive Ringer solution. *A* represents data from a whole nerve preparation; *B* shows the results obtained with a bundle of seven 30 μ fibres. (After Keynes, 1951.)

The exchange of divalent cations

The escape of ^{45}Ca from the isolated nerve cord of *Periplaneta* (Treherne, 1962 b) and, in addition, that of ^{28}Mg from this preparation in *Carausius* (Treherne, 1965 b) was found to occur as a two-stage process. By analogy with the behaviour of sodium ions the rapidly exchanging components were identified as the extracellular and the slowly exchanging ones as the intracellular ion fractions. The movements of calcium ions between the extracellular fluid and the external medium have been shown to be lower than would be expected by comparison with their free diffusion constants (fig. 10). As has already been suggested for the monovalent cations such an effect could result from the attraction of fixed anion groups within the extracellular system or from the presence of peripheral fixed cationic groups.

37

The level of the exchangeable extracellular calcium and magnesium has been estimated from the kinetics of ^{41}Ca and ^{28}Mg exchange in the central nervous systems of the above species. The calculated extracellular concentrations of these ions are given in tables 4 and 5. The extracellular concentration of calcium in the nerve cord of *Carausius* was not affected by the presence of metabolic inhibitors (Treherne, 1965b), and with both species it will be seen that the relatively high concentration of the divalent cations in this fluid appeared to be governed by a Donnan equilibrium with the external medium. In these investigations the *in vitro* use of appropriate physiological salines avoided any complications due to the presence of appreciable amounts of 'bound' calcium or magnesium in the haemolymph. It seems most likely, as has already been discussed with regard to the monovalent cations, that the presence of fixed anion groups associated with extracellular mucopolysaccharides (Ashhurst, 1961b; Pipa, 1961) or with collagen-like material (Gray, 1960) might contribute to the demonstrated Donnan equilibrium with the blood.

The above data, although demonstrating the existence of an appreciable exchangeable fraction, do not necessarily indicate the total level of the extracellular divalent cations, for it is possible that significant amounts of inexchangeable or very slowly exchanging calcium and magnesium might be present (Shanes, 1964). Although the electrical properties of the nerve cells are most likely to be related to the exchangeable ions, the presence of 'bound' calcium or magnesium would be of importance in the estimation of their intracellular concentrations. In both crustacean (Lewis, 1952) and insect nerves (Treherne, 1962b, 1965b) the intracellular ion concentrations were arrived at by subtraction of the exchangeable extracellular fraction from the total tissue content determined by chemical analysis. The intracellular values for magnesium and calcium contained in tables 4 and 6 must, therefore, be regarded as maximum estimates of their concentrations.

By analogy with the cells of other tissues (cf. Shanes, 1964) it is likely that a large proportion of the intracellular calcium and magnesium may be inexchangeable and that, as in squid axoplasm in which calcium ions move at less than 1/30 the rate in free solution (Hodgkin and Keynes, 1957), their activity may be much reduced. Rudenberg (1954) has reported that approximately half of the calcium in homogenates of whole lobster nerve is only slowly removed by dialysis against distilled water. The fact that a large proportion of the axoplasm calcium may not be in a freely ionized state probably results from the tendency of this ion to combine with such phosphorus compounds as ATP, phospholipids and nucleic acids, which, it has been pointed out (Keynes and Lewis, 1956), are all present in crab nerve in relatively high concentrations. In view of the fairly large exchangeable extracellular fractions of the divalent cations (Treherne, 1962b, 1965b), it is likely that the activity gradients between the extracellular fluid and the axoplasm are greater than the gradients of chemical concentrations between the two phases. This is of particular interest in the case of magnesium in *Carausius* nerve cord, in which the high exchangeable extracellular concentration of this ion already exceeds the apparent intracellular level by a factor of approximately 11·0 (table 6).

An attempt to measure the intracellular calcium content of *Carcinus* peripheral nerve was made by washing away the extracellular ions by soaking in calcium-free physiological solution for 10–15 min (Keynes and Lewis, 1956). Despite the possibility of errors, due to incompleteness of extracellular exchange or the escape of intracellular calcium during the soaking period, the calculated figure of 0·54 mM/kg is in good agreement with the values obtained by direct analysis of squid axoplasm. The very much higher values for the calculated intracellular concentrations in insects obtained from the study of the kinetics of ^{41}Ca exchange (Treherne, 1962b; 1965b) may be a reflexion of a larger proportion of un-ionized calcium in the nerve cells of these species.

The escape of ^{41}Ca and ^{28}Mg from the nerve cord of *Carausius* occurred with measured half-times of 890·0 and 720·0 sec (Treherne, 1965b), which correspond to exchange constants (K_{out}) of $7·78 \times 10^{-4}$ and $9·61 \times 10^{-4}$ sec^{-1} respectively. There is a possible objection to the calculation of the inward transfer constant (K_{in}), in view of the possibility of a proportion of inexchangeable ions in the intracellular concentrations shown in table 6. If, however, the intracellular activity coefficients of these ions were approximately the same then it follows, from the very high extracellular level of magnesium, that there is a greater inward permeability to calcium than to magnesium ions in the cells of the central nervous system of this insect.

The exchanges of labelled chloride ions between the extra-cellular fluid and the external medium were found to occur relatively rapidly in the nerve cords of both the species of insects which have been investigated (Treherne, 1962 b, 1965 b). The apparent movements of these anions in the extracellular system of the cockroach nerve cord approximated to those of tritiated water and non-electrolytes and there was clearly no restriction to diffusion of the type encountered by the mono-valent and divalent cations (fig. 10).

The concentration of the rapidly exchanging extracellular chloride in the central nervous tissues of *Periplaneta* was esti-mated to be considerably lower than in the external medium, owing to the presence of fixed or indiffusible anion groups in the spaces between the cells (table 4). The difficulties encountered in defining the rapidly exchanging fraction in the nerve cord of *Carausius* make it impossible to compare the extracellular level of this ion with the concentration in the external medium (table 5). The initial uptake of ^{36}Cl in the nerve cord of this species was reduced in the presence of metabolic inhibitors (Treherne, 1965 b). It seems most likely that a significant pro-portion of the influx of chloride into the extracellular fluid may occur in association with sodium ions, which have been shown to be actively transported from the external medium in *Carausius*.

The efflux of ^{36}Cl from the intracellular fraction of the cock-roach nerve cord exhibited a half-time of 126·0 sec (Treherne, 1962 b), which corresponds to an exchange constant (K_{out}) of $5·5 \times 10^{-3}$ sec^{-1}. The above value is appreciably greater than that of $2·5 \times 10^{-3}$ sec^{-1} calculated for sodium (Treherne, 1961 e) or $0·75 \times 10^{-3}$ sec^{-1} for potassium efflux (Treherne, 1962 b). It seems clear, therefore, that the movements of this anion between the intracellular and the extracellular fluid are taking place con-siderably more rapidly than the equivalent exchanges of cations.

The intracellular concentration of chloride ions has been estimated in peripheral nerve of the crab *Carcinus maenas* (Lewis, 1952) (table 3). The calculated value of 27·0 mM/kg water can, as has already been discussed in this chapter, be related to the presence of indiffusible anion groups, resulting in a Donnan equilibrium with the extracellular fluid in which the chloride level was assumed to be that of external medium (145 mM/l).

CHAPTER 4

The Ionic Basis of Electrical Activity

THE preceding chapter summarized our existing knowledge of the distribution and exchange of the common inorganic ions in arthropod nervous tissues. It was shown that, despite the peculiarities of the blood in certain arthropod species, the ionic composition of the extracellular fluid approximated to that of the body fluids in most animal groups. The present chapter represents an attempt to interpret these results in terms of some current concepts of the mechanism of nervous transmission. Consideration will also be given to some structural peculiarities of arthropod nervous tissues which have been shown to influence the ionic events occurring during electrical activity.

The effects of inorganic ions on whole nerves and ganglia

Although fully quantitative analyses on the chemistry of nervous transmission could only be carried out with the development of satisfactory intracellular micro-electrode techniques, some useful information has emerged from studies on the effects of ions on whole nerves and ganglia using external electrodes. It has long been recognized that increase in the potassium concentration of the external solution causes a decrease in the injury potential of crustacean peripheral nerve (Cowan, 1934; Shanes and Hopkins, 1948). At high potassium levels the injury potential of *Maia* nerve was found to vary as the logarithm of the external concentration of the ion (Cowan, 1934). The 'spontaneous' activity of whole abdominal ganglia of the crayfish has also been shown to be enhanced at low levels and decreased at high external potassium concentrations (Prosser, 1940a, 1943). In whole crab nerve, calcium depletion of the medium lowers the maximum injury potential which can be attained with low external potassium levels (Shanes and Hopkins, 1948). The effects of elevated potassium concentrations in causing depolarization can be reduced with increased levels of multivalent cations in spider crab nerve (Guttman, 1939), although the effect of increased calcium was only slight in the cockroach nerve cord (Treherne, 1962 b).

The nerves of insects appear to differ from those of crustaceans in the ability to continue functioning in the presence of appreciable concentrations of potassium in the external medium.

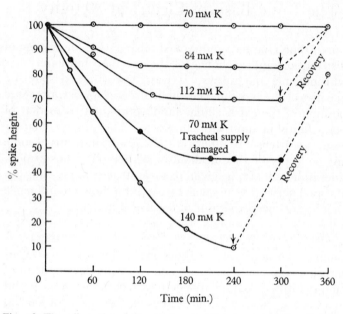

Fig. 16. The effect of variation in the potassium concentration of the bathing medium upon the spike height of a single motor axon of the crural nerve of the locust *Locusta migratoria*. Recovery following bathing in a 10·0 mM/l potassium solution is indicated by the arrows and broken lines. (After Hoyle, 1953.)

The crural nerve of *Locusta migratoria migratorioides* was unaffected by the presence of 70·0 mM/l potassium for periods of several hours (Hoyle, 1953). Only relatively slow depolarization was obtained with even high external potassium concentrations, although damage to the nerve tracheal supply caused a rapid loss of function when the nerve was bathed with the 70·0 mM/l solution (fig. 16). Injection of high potassium solutions beneath the nerve sheath caused a rapid depolarization. The presence of high external concentrations of potassium caused a more rapid loss of conduction in the abdominal nerve cord of the cockroach, although in this species also there was an enhanced rate of depolarization in the absence of the fibrous and cellular nerve sheath (Twarog and Roeder, 1956). Such effects might be expected if the peripheral nerve sheath functioned as an effective diffusion barrier restricting the entry of ions and molecules into the underlying tissues. As has already been described in the previous chapter, however, the results of radioisotope experiments showed that the exchanges of inorganic ions and molecules occur relatively rapidly between the blood and the central nervous system in both the cockroach (Treherne, 1961 *a*, *c*, *e*; 1962 *a*, *b*) and the stick insect, *Carausius morosus* (Treherne, 1965 *b*). With the cockroach these exchanges were shown to be those taking place with the extracellular fluid, in which the high levels of cations and low concentrations of chloride ions appeared to be caused by a Donnan equilibrium with the blood (Treherne, 1962 *b*). Removal of the nerve sheath caused some very dramatic changes in the ionic composition of the extracellular fluid, due to the disruption of the Donnan equilibrium with the blood, and also a sixfold increase in the measured inulin space of the central nervous tissues. The enhanced rates of depolarization in the absence of the nerve sheath in the central nervous system of *Periplaneta* might, therefore, be a reflexion of the changed ionic environment rather than to any properties of the sheath as a diffusion barrier. It was shown, in fact, that the rate of decrease of compound action potentials in desheathed preparations occurred more slowly at elevated potassium concentrations when the concentrations of the major cations were raised to approximate to the composition of the extracellular fluid of the intact nerve cord under these conditions (fig. 17) (Treherne, 1962 *c*). The delayed potassium depolarization in these experiments appeared to be due to the increased concentration of sodium rather than to the calcium ions. This effect appears to be analogous to that demonstrated in frog nerve fibres by Lundberg (1951). The delay in depolarization due to the interaction of the other extracellular cations is, however, unlikely to be the only effect causing differences between intact and desheathed cockroach nerve cords. It seems probable that there will be an increased accessibility to potassium in the central nervous tissues in desheathed preparations, resulting from the changes reflected in the very dramatic increase in the measured inulin space.

It has been pointed out (Smith and Treherne, 1963) that the

above considerations do not necessarily apply to the peripheral nerve of the locust, in which the rate of potassium depolarization was an order slower than in the intact cockroach nerve cord (Hoyle, 1953). There is, in fact, some evidence of apparent differences in accessibility of sodium ions between the peripheral and central nervous tissues of *Carausius morosus* (Wood, 1957; Treherne, 1965*a*).

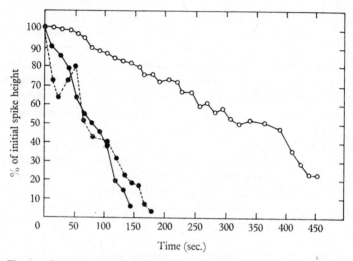

Fig. 17. Comparison of the rate of loss of conduction across the desheathed fourth abdominal ganglion of the cockroach *Periplaneta americana* when perfused with a high potassium solution (70 mM/l K$^+$; 99·3 mM/l Na$^+$; 4·5 mM/l Ca^{++}) (closed circles) and with a solution containing elevated cation concentrations (open circles) corresponding to the levels in the extracellular fluid of prefused intact ganglia (85·9 mM/l K$^+$; 153·0 mM/l Na$^+$; 17·1 mM/l Ca^{++}). These observations were carried out on the same preparation, which was washed with normal physiological solution between each perfusion to restore the original electrical activity. The preparation was first perfused with the 70 mM/l K$^+$ solution (continuous line), then with the 'extracellular solution', and then once again with the 70 mM/l K$^+$ solution. (From Treherne, 1962*c*.)

The relation between excitability and the external sodium concentration has been known since the original investigations by Overton (1902). Crustacean peripheral nerve was found to become rapidly inexcitable when the external sodium was replaced by choline or tetraethylammonium ions, the effect being immediately reversed by washing the preparation in normal Ringer solution (Burke, Katz and Machne, 1953). The intact cockroach nerve cord showed only a slow loss of conduction in sodium-deficient solution, although rapid blockage occurred when the preparation was desheathed (Twarog and Roeder, 1956). The very rapid and apparently passive exchanges of sodium ions across the nerve sheath in this species suggests that this effect might result from some local regulation of the extracellular fluid in the immediate vicinity of the axon surface in the intact nerve cord. It has already been mentioned in the previous chapter that there is regulation of an appreciable proportion of the extracellular sodium in the nerve cord of the stick insect (Treherne, 1965b). This observation supplies an explanation of the effect of the removal of the very low level of external sodium ions which produced a relatively rapid loss of conduction in both intact and desheathed preparations (Treherne, 1965a). The intact peripheral nerve in this species is, however, able to function for extended periods in the absence of external sodium ions (Wood, 1957). As has already been suggested with respect to potassium, this effect could result from differences in accessibility of cations between the peripheral and central nervous systems of the phytophagous insects.

There is relatively little information on the effects of cations other than sodium and potassium on the electrical behaviour of whole nerves and ganglia. Substitution of lithium had little effect on peripheral nerve of *Maia*, while substitution of rubidium and caesium caused only small changes to the injury potential, although these latter ions were found to cause a reversible abolition of the action current (Cowan, 1934). The alkaline earths (Ba, Sr, Ca and Mg) had no appreciable effect on the injury potential of peripheral nerve in this species (Guttman, 1939), a conclusion which contrasts with later investigations on the effects of calcium ions using intracellular electrode techniques. Elevated calcium concentration was, however, found to reduce the rate of potassium depolarization in this crustacean. Calcium-lack has been shown to cause spontaneous firing of multifibre preparations of lobster nerve (Gordon and Welsh, 1948). In the phytophagous insect *Carausius morosus*, in which magnesium occurs in relatively high concentrations in the blood (table 2), there was a loss of conduction in the abdominal nerve cord when this ion was removed from the bathing solution (Treherne, 1965a).

There have been few studies on the effects of extracellular monovalent anions on the electrical activity of nerves and

45

ganglia in arthropods. Some minor increases in injury potential were obtained in nerves of *Maia squinado* in experiments involving substitution of inorganic and organic anions (Wilbrandt, 1937).

The resting potential

The data on the probable distribution of inorganic ions in arthropod nervous tissues suggest that the relatively high internal potassium level and the low concentration of chloride can be attributed as a first approximation to a Donnan equilibrium with the extracellular fluid, as originally proposed in the classical experiments of Boyle and Conway (1941). As with excitable tissues from other groups of organisms, the low intracellular sodium does not appear to be related to any degree of impermeability of the cell membrane, but, as has been demonstrated in insect axons, to an active extrusion of the ion from the axoplasm. It was also suggested in the previous chapter that the uptake of potassium by a linked ion pump (which has been demonstrated in some arthropod nerves) might also result in some departure from a simple Donnan equilibrium with the extracellular fluids.

According to the theory of Boyle and Conway the resting potential (E_m) of a steady-state axon can, as a first approximation, be related to the equilibrium potentials for potassium and chloride $(E_K$ and $E_{Cl})$ by the Nernst equations:

$$E_m = E_K = E_{Cl},$$

$$E_K = \frac{RT}{F} \log_e \frac{[K]_i}{[K]_o},$$

$$E_{Cl} = \frac{RT}{F} \log_e \frac{[Cl]_o}{[Cl]_i},$$

where R is the gas constant; T the absolute temperature; F the Faraday; $[K]_o$, $[K]_i$, $[Cl]_o$ and $[Cl]_i$ the activities of the ions inside and outside the fibre. The activity coefficient of potassium in squid axoplasm at least is known to be similar to that in free solution (Hinke, 1961), so that it is probably valid to derive the potassium equilibrium potentials from a comparison of the chemical concentrations of the ion in the external solution and within the axons in arthropods.

The measured resting potentials together with the calculated

46

potassium equilibrium potentials for some arthropod axons are shown in table 7. The equilibrium potentials shown in table 7 were calculated from intracellular concentrations estimated either (as in the case of the insect species) from the kinetics of potassium exchange or (as in the crustacean nerves) from measurements based on the observed polysaccharide space.

TABLE 7. *The measured resting potential* (E_m) *and the potassium equilibrium potential* (E_K), *calculated from the intracellular* $([K]_i)$ *and extracellular* $([K]_o)$ *potassium levels, for some arthropod nerve cells. These values are compared with those obtained for squid giant axons*

Species	Preparation	E_m (mV)	E_K (mV)	$[K]_i$ (mM/l)	$[K]_o$ (mM/l)	Authorities
CRUSTACEA						
Homarus americanus	Ventral nerve cord	73·3[1] and 71·0[2]	86·2	307·1[3]*	10·0	[1]Tobias and Bryant (1955) [2]Dalton (1958) [3]Fenn *et al.* (1934)
Homarus vulgaris	Peripheral nerve	62·0[1]	—	—	10·0	[1]Hodgkin and Huxley (1945)
Orconectes virilis	Ventral nerve cord	85·0[1]	—	—	5·4	[1]Dalton (1959)
Procambarus alleni	Nerve cell of stretch receptor	61·0[1] and 70–80·0[2]	—	—	5·4	[1]Edwards *et al.* (1963) [2]Eyzaguirre and Kuffler (1955)
Carcinus maenas	Peripheral nerve	71·0–94·0[1]	91·2–92·5	412–432·0[2]	10·0	[1]Hodgkin and Huxley (1945) [2]Lewis (1952)
INSECTA						
Periplaneta americana	Nerve cord	77·0[1] 57·0[2]	65·7	225·1[3]	3·1 17·0[3]	[1]Narahashi and Yamasaki (1960) [2]Yamasaki and Narahashi (1959*a*) [3]Treherne (1962*b*)
Carausius morosus	Nerve cord	—	37·7	555·8[1]	124·5[1]	[1]Treherne (1965*b*)
CEPHALOPODA						
Loligo pealii	Giant axon	77·0[1]	78·1	369·0[2]	16·6[3]	[1]Moore and Cole (1960) [2]Steinbach and Spiegelman (1943) [3]Manery (1939)

* Calculated on the basis of an extracellular space of 0·242 l/kg tissue (Lewis, 1952) and a water content of 89·2 % (Nevis, 1958).

These values differ from previous estimates for crustacean (Hodgkin, 1951; Shanes, 1958a) and insect axons (Narahashi, 1963) in which potassium concentrations for whole peripheral nerves or nerve cords were used. In three species only is it possible to compare directly the two potentials. In the axons of the crab, lobster and an insect in which this comparison is possible the potassium equilibrium potentials calculated from the Nernst equation approach the measured resting potentials, but in none of these arthropods is the agreement as close as the values for the cephalopod giant axon shown in table 7. The latter values are at the moment probably unique in that the resting potentials were derived from *in vivo* experiments and the internal potassium concentration was measured by direct analysis of the axoplasm. The values for the arthropod axons summarized in table 7 show, however, a similar relation to those obtained in earlier measurements on squid axons and other excitable tissues (Hodgkin, 1951; Shanes, 1958a). It therefore seems reasonable to conclude that resting potential can approximately be described in terms of the concentration gradient for potassium ions across the axon membrane as in cephalopods.

The calculated potassium equilibrium potential for the phytophagous insect *Carausius morosus* shown in table 7 is of interest in view of the peculiar ionic balance of the blood, in which the potassium level exceeds that of sodium (table 2). In spite of the very high extracellular concentration of potassium in this species the calculated equilibrium potential is approximately 37·7 mV because of the exceedingly high intracellular concentration of this ion (Treherne, 1965b). This potassium equilibrium potential is, however, close to the low observed resting potential of 41·0 mV for muscle fibres of this insect (Wood, 1957). The estimated equilibrium potential should in any case be regarded as a minimum estimate in view of the possibility (which has already been discussed in the previous chapter) that the activity coefficient of the extracellular potassium may be lower than that in free solution.

Part of the discrepancy between the calculated potassium equilibrium potentials and the resting potentials summarized in table 7 can be attributed to the fact that, as has been shown for *Carcinus* nerve (Keynes and Lewis, 1951), the isolated axons are permeable to other ions than potassium. For the case where the electrical field is constant throughout the membrane the equation defining the membrane potential is given by the equation developed by Hodgkin and Katz (1949):

$$E_m = \frac{RT}{F} \log_e \frac{P_K[K]_i + P_{Na}[Na]_i + P_{Cl}[Cl]_o}{P_K[K]_o + P_{Na}[Na]_o + P_{Cl}[Cl]_i},$$

where P_K, P_{Na} and P_{Cl} are the relative permeabilities to the various ions.

The action potential

Action potentials have been recorded with intracellular micro-electrodes in the nerve cells of several crustacean and one insect species (fig. 18 and table 8). In all cases there is a reversal of the membrane potential in these arthropod axons which varies between 22·0 and 52·0 mV. According to the membrane theory of the propagation of the action potential, this overshoot is supposed to result from a transient increase in sodium conductance, the crest of the spike approaching the equilibrium potential for sodium given by the Nernst equation:

$$E_{Na} = \frac{RT}{F} \log_e \frac{[Na]_o}{[Na]_i},$$

where $[Na]_o$ and $[Na]_i$ are the sodium concentrations outside and inside the axon. Strictly speaking, the sodium activities should be used in this equation rather than the values for the chemical concentrations of the ion. In squid axoplasm it has been shown, in fact, that the activity of the sodium ions is somewhat lower than in free solution (Hinke, 1961), so that values shown for the sodium equilibrium potential in table 8 should be regarded as minimum estimates.

A comparison of the sodium equilibrium potential with the reversed potential is possible for only two species of arthropods, the crab *Carcinus maenas* and the insect *Periplaneta americana* (table 8). In both cases the observed potentials fall short of the estimated equilibrium potentials, a result which would be expected if the active membranes were not completely selective for sodium ions (Hodgkin, 1951, 1964).

It is of some interest to note that the calculated sodium equilibrium potential for the axons of the phytophagous insect *Carausius morosus* (in which the blood level of this cation is very low) is of the same order of magnitude as that for the cockroach, owing to the active maintenance of a high extracellular sodium

4

TABLE 8. *The magnitude of the resting (E_m) and action potentials (E_a) for some arthropod nerve cells, compared with the sodium equilibrium potential (E_{Na}) calculated from the intracellular $[Na]_i$ and external sodium concentrations $[Na]_o$*

Species	Preparation	E_m (mV)	E_m-E_a (mV)	E_a (mV)	E_{Na} (mV)	$[Na]_i$ (mM/l)	$[Na]_o$ (mM/l)	Authorities
CRUSTACEA								
Homarus americanus	Ventral nerve cord	73·3[1] and 71·0[2]	100·7[1] and 117·9[2]	−27·4 and −46·9[2]	—	—	465·0	[1]Tobias and Bryant (1955)
							465·0	[2]Dalton (1958)
Homarus vulgaris	Peripheral nerve	62·0[1]	106·0[1]	−44·0	—	—	460·0	[1]Hodgkin and Huxley (1945)
Orconectes virilis	Ventral nerve cord	85·0[1]	114·0[1]	−29·0	—	—	205·3	[1]Dalton (1959)
Procambarus alleni	Nerve cell of stretch receptor	61·0[1] and 70–80·0[2]	77·1[1] and 80–100·0[2]	−16·1– 20·0	—	—	205·3	[1]Edwards et al. (1963)
							205·3	[2]Eyzaguirre and Kuffler (1955)
Carcinus maenas	Peripheral nerve	71·0–94·0[1]	134·0[1]	−52·0	−54·4– 68·7	30–53[2]	460·0	[1]Hodgkin and Huxley (1945)
								[2]Lewis (1952)
INSECTA								
Periplaneta americana	Nerve cord	77·0[1]	99·0[1]	−22·0	−29·1	67·2[2]	214·0	[1]Narahashi and Yamasaki (1960)
								[2]Treherne (1962b)
Carausius morosus	Nerve cord	—	—	—	−22·3	86·3[1]	212·4[1]	[1]Treherne (1965b)
CEPHALOPODA								
Loligo pealii	Giant axon	77·0[1]	118·0[1]	−41·0	−52·5	44·0[2]	354·0[3]	[1]Moore and Cole (1960)
								[2]Steinbach and Spiegelman (1943)
								[3]Manery (1939)

concentration in the nervous tissues of the former species (Treherne, 1965 b).

There has been no attempt to measure directly the amount of sodium entering the axons of arthropods during activity, although estimates have been made for the associated escape of potassium ions during the falling phase of the action potential (Cowan, 1934; Young, 1938; Keynes, 1951; Abbott, 1958). These estimates have been discussed in the previous chapter (p. 33). It is, however, possible to estimate the sodium entry

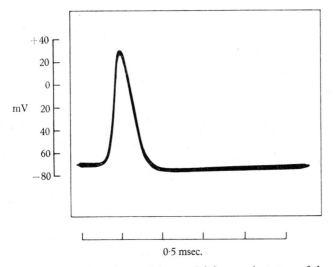

Fig. 18. Recording of an action potential from a giant axon of the abdominal nerve cord of the cockroach *Periplaneta americana*. (From Narahashi, 1963.)

during the propagation of an action potential indirectly from the values of the membrane electrical constants. According to Hodgkin and Katz (1949) the total quantity of sodium entering the axon during excitation is given by CV/F, where C is the membrane capacity, V the action potential and F the Faraday. These constants are known for two crustacean and one insect species and can be used to calculate the sodium entry into the axon per impulse (table 9).

The calculated values shown in table 9 are likely to be minimum estimates of the sodium entry, for, as Hodgkin (1951) has pointed out, the sodium entry during the rising phase may be partly neutralized by the exit of potassium or the entry of chloride ions. It is also possible that sodium entry may continue during the falling phase.

TABLE 9. *The total amount of sodium* (I_{Na}) *entering some arthropod axons during excitation, calculated from membrane capacity* (C) *and the action potential* (V)

Species	C (μF/cm^2)	V (mV)	I_{Na} [($\times$ 10^{-12} M/cm^2/ impulse)]
Carcinus maenas	1·1 (Hodgkin, 1947)	134·0 (Hodgkin and Huxley, 1945)	1·5
Homarus vulgaris	1·3 (Hodgkin and Rushton, 1946)	106·0 (Hodgkin and Huxley, 1945)	1·4
Periplaneta americana	6·3 (Yamasaki and Narahashi, 1959*b*)	99·0 (Narahashi and Yamasaki, 1960)	6·4

The calculated figure of 1·5 $\times$ 10^{-12} Mcm^{-2} impulse^{-1} for the inward movement of sodium in the axons of *Carcinus* thus corresponds well with the figure of 1·7 $\times$ 10^{-12} Mcm^{-2} impulse^{-1} for the escape of potassium estimated by indirect methods (Hodgkin and Huxley, 1947). The former figure is also of a similar order of magnitude to that of 2·4 $\times$ 10^{-12} cm^{-2} impulse^{-1} obtained from direct measurements using ^{42}K in the axons of this species (Keynes, 1951). There is thus reasonable agreement between the quantity of sodium which enters and the amount of potassium which leaves the axon during a single impulse in this crustacean. This accords with the membrane theory of the action potential, in which it would be expected that the charge transferred into the axon during the rising phase would be balanced by an equivalent outflow during the subsequent falling phase.

In *Periplaneta* giant axons the unusually high membrane capacity of 6·3 μF/cm^2 (Yamasaki and Narahashi, 1959*b*) yields a minimum calculated sodium influx of 6·4 $\times$ 10^{12} Mcm^{-2} impulse^{-1} which is much larger than in the crustacean species shown in table 9. This membrane capacity is equivalent to that of the frog nerve fibre (Fatt and Katz, 1951) in which, nevertheless, the sodium entry in stimulated preparations (Hodgkin and Horowicz, 1959) was found to be sufficient to generate the electrical change in the active membrane (Hodgkin, 1964).

The effects of variations in the external potassium concentration on the resting and action potentials

In the previous sections the resting and action potentials of some arthropod axons have been described in terms of sodium and potassium conductances and the concentration gradients of these ions between the extracellular fluid and the axoplasm. It is now relevant to consider the available evidence concerning the effects of variation in the external ionic concentrations on these potentials as measured with intracellular electrodes.

As has already been mentioned with respect to multifibre preparations, the effect of increased potassium in the external solution is to decrease the measured injury potential. This effect has been confirmed, using intracellular electrodes, when it was shown that there is a depolarization with increased potassium concentration in the giant axons of the lobster *Homarus americanus* (Dalton, 1958), the crayfish *Orconectes virilis* (Dalton, 1959) and the cockroach *Periplaneta americana* (Yamasaki and Narahashi, 1959a), and the neurones of the abdominal stretch receptors of two species of crayfish (Edwards, Terzuolo and Washizu, 1963). Figure 19 illustrates the effect of external potassium ions on the resting potential of cockroach giant axons over a wide range of concentrations (Yamasaki and Narahashi, 1959a). In this insect decrease in the external potassium level has very little effect on the resting potential. At concentrations above 20 mM/l the resting potential showed a decline which can be represented as a straight line on the logarithmic scale. Now from the Nernst equation it would be expected that over the straight portion of the graph the resting potential would change by approximately 58·0 mV for a tenfold alteration in potassium concentration. The slope of the line illustrated in fig. 19 actually represents a 42·0 mV change for a tenfold concentration increase. A similar departure from that predicted by the Nernst equation was also noted in the giant axons of crayfish (fig. 20) and the lobster (Dalton, 1958, 1959) and the crustacean stretch receptor neurone (Edwards *et al.*, 1963). These observations imply that the conductance to ions other than potassium must contribute to the resting potential of the axons in these arthropod

53

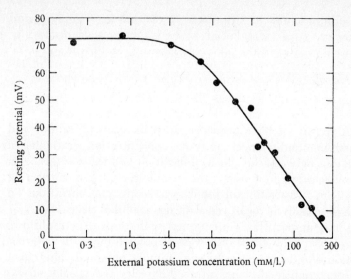

Fig. 19. The effect of variation of the external potassium concentration on the resting potential in the cockroach giant axon. (After Yamasaki and Narahashi, 1959*a*.)

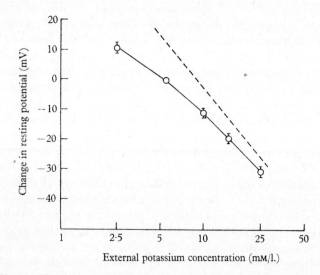

Fig. 20. The change in magnitude of the resting potential of crayfish giant axons produced as a result of alterations in the potassium concentration of the bathing medium. The broken line represents the change in the potassium equilibrium potential predicted by the Nernst equation. (After Dalton, 1959.)

species. Such an effect could cause the discrepancy between the potassium equilibrium potentials and the observed resting potentials recorded in table 7. This state of affairs appears to be essentially similar to that encountered in similar experiments on the squid giant axon (Curtis and Cole, 1942), where, allowing for the permeabilities of potassium, sodium, chloride and the variations with potential, the resting potential can be well accounted for (Baker, Hodgkin and Shaw, 1962).

Quantitative studies on *Homarus* giant axons have shown that in conditions of high external potassium the magnitude of the action potential is dependent on that of the resting potential, the action potential amplitude being proportional to the logarithm of the resting potential minus a constant (Dalton and Adelman, 1960). There was, however, no evidence of a direct effect of the potassium ions on the size of the overshoot in this species.

The effects of variation of the external sodium concentration on the resting and action potentials

Alterations in the external concentration of sodium ions has been shown to produce little change in the magnitude of the resting potential of the insect axon (fig. 21) (Yamasaki and Narahashi, 1959a; Boistel, 1960). However, as predicted by the membrane theory and as has been shown with external electrodes in multi-fibre preparations, the action potential showed a striking dependence on the external sodium concentration in the giant axons of species of lobster and crayfish (Dalton, 1958, 1959) and of the cockroach (fig. 21) (Yamasaki and Narahashi, 1959a).

The alternations in magnitude of the action potential with varying external concentrations of sodium ions appear to bear a close relation to the 58·0 mV change per tenfold increase in concentration (which would be expected from the Nernst equation for the sodium equilibrium potential) in crayfish giant axons (fig. 22) (Dalton, 1959). In this case it is, therefore, reasonable to suppose that the peak value of the action potential is largely determined by the high conductance of sodium ions. In cockroach giant axons, however, there is some departure of the measured action potentials from the 58·0 mV slope which would be expected if the axon membrane behaved as a perfect sodium electrode (fig. 21). In this species, therefore, it is necessary to conclude that the conductances of other ions participate in determining the peak of the action potential (Yamasaki and Narahashi, 1959a; Narahashi, 1963). This effect would also accord with the difference observed between the measured action potential and the sodium equilibrium potential (table 8).

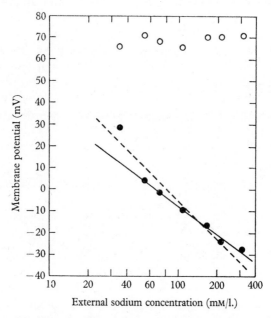

Fig. 21. The effect of variation of the external sodium concentration upon the resting potential (open circles) and active membrane potential (closed circles) in cockroach giant axons. The broken line represents the sodium equilibrium potential calculated according to the Nernst equation. The solid line is drawn through the observed values, excluding the lowest one. (After Yamasaki and Narahashi, 1959a.)

The effects of lithium on resting and action potentials

As in other excitable tissues it has been found that substitution of lithium for sodium ions in the external solution produced no apparent change in the resting or action potentials of cockroach giant axons (Narahashi, 1963).

The effects of calcium on resting and action potentials

It has long been known that the presence of calcium ions is essential for the normal functioning of various excitable tissues (cf. Brink, 1954; Shanes, 1958a, b). The study of the effects of variations in the concentration of calcium ions is of especial interest in the Crustacea in view of the very wide fluctuations in the calcium level of the blood (Cole, 1941), which is probably related to the moulting cycle (Travis, 1955). Among the insects the relatively high extracellular level of the exchangeable calcium (Treherne, 1962 *b*) might also be expected to influence the electrical activity of the axons.

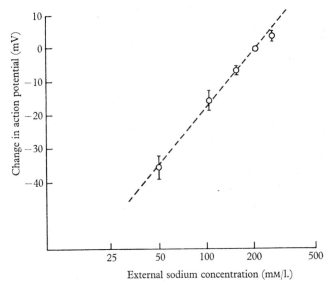

Fig. 22. The effect of alteration of the external sodium concentration on the recorded action potentials in giant fibres of the crayfish *Orconectes virilis*. The broken line, which has a slope of 58 mV per tenfold change in sodium concentration, shows the change in the sodium equilibrium potential predicted by the Nernst equation. (After Dalton, 1959.)

The reduction of external calcium concentration is well known to produce spontaneous discharge or repetitive firing in response to a single shock in a variety of excitable tissues (cf. Brink, 1954). Such an effect was demonstrated in multi-fibre preparations (Gorden and Welsh, 1948) and in single lobster motor axons, which became spontaneously active in calcium-free solution, firing at frequencies as rapid as 100/sec (Adelman, 1956). The latter activity declined rapidly, however, both spike discharge and repetitive firing eventually ceasing altogether. A slow spontaneous type of repetitive activity, with a frequency of around 1/sec, was observed with calcium-free solution in the giant axons of this species (Dalton, 1958). No spontaneous discharge or repetitive firing in calcium-free solution was noted in the giant axons of either the crayfish (Dalton, 1959) or the cockroach (Narahashi and Yamasaki, 1960). It is not known to what extent traces of calcium adhering to the axonal membrane or the extracellular material may prevent the appearance of spontaneous effects in the latter species, as decalcifying agents were not used in these experiments.

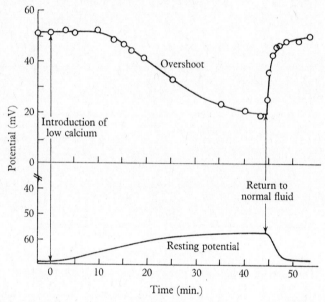

Fig. 23. Time course of the potential changes in the giant axons of the lobster *Homarus americanus* following a change of the bathing medium from a normal calcium level (25 mM/l) to one containing zero calcium. The distance between the two curves shown in this diagram represents the extent of the action potential at any given time during the experiment. (After Dalton, 1958.)

Decreases in the measured resting and action potentials have been noted with low external calcium concentrations in giant and motor axons of the lobster (fig. 23) (Dalton, 1958; Adelman and Adams, 1959), and in the giant axons of the crayfish (Dalton, 1959) and the cockroach (Narahashi and Yamasaki, 1960). In the lobster and the cockroach, but not the crayfish, changes in the external calcium concentration produced marked effects on the form of the action potential. In both the former species the duration of the action potentials was decreased at lowered calcium concentrations and increased at elevated levels, owing to changes in the rate of rise and to alternations in the onset and rate of decline of the action potentials. In the cockroach complete removal of the external calcium caused depolarization and development of a conduction block in the giant axons.

The effects of the extracellular concentration of calcium ions on the rate of potassium depolarization in multi-fibre preparations have already been commented on (Guttman, 1939). This effect is not necessarily universal in arthropods, for elevated calcium concentration has been found to be without influence on potassium depolarization in the neurone of the crayfish stretch receptor, although it was possible to demonstrate appreciable effects of calcium-potassium antagonism on the membrane resistance of this preparation (Edwards et al., 1963).

The results of experiments on lobster giant axons suggest that external calcium plays an important part in the spike-generating mechanism, which is independent of the effects caused by alterations in the resting potential produced by this ion (Dalton and Adelman, 1960). Further experiments also suggested an interaction of calcium with sodium and potassium during the spike (Adelman and Dalton, 1960) which could be accounted for by changes in the sodium and potassium conductances as demonstrated by the voltage-clamp technique in squid giant axons (Frankenhaeuser and Hodgkin, 1957). More recently the results of experiments on repetitive stimulation have demonstrated the role of calcium in governing the permeability of crustacean axon membranes to sodium and potassium ions (Wright and Tomita, 1965). Finally, an effect of calcium on sodium movements can be inferred from the observation that the excitability of tetrodotoxin-poisoned lobster axons, in which the sodium-carrying system is thought to be suppressed (Narahashi, Moore and Scott, 1964), can be partly restored by anodal depolarization when the external concentration of calcium ions is reduced (Narahashi, 1964).

Changes in the permeability of axon membranes to monovalent cations due to the addition or removal of calcium from sites or carriers within the membrane was postulated in the theory advanced by Gordon and Welsh (1948) to account for their observations on the effects of this ion on lobster axons. According to such a theory the movement of sodium ions would depend upon the presence of calcium ions within pores in the

59

membrane, removal of which would allow an influx of the monovalent cations to take place through the pores. Such a theory would demand that, as depolarization is held to remove calcium from the membrane, the sites occupied by the ions would be near the inner edge of the membrane and would be accessible to external but not internal calcium (cf. Hodgkin, 1964). The theory of Gordon and Welsh proved inadequate in the quantitative prediction of the resting potential changes produced by alterations in the external calcium level (Frankenhaeuser and Hodgkin, 1957). The discrepancy between the above theory and these later observations could be accounted for by the assumption that several calcium ions are involved with an inorganic anion in blocking membrane pores, or that the calcium ions are only adsorbed on the periphery of the membrane in such a way that they alter the electrical field within it, without affecting the overall potential difference between the axoplasm and the external solution (Hodgkin, 1964). The mechanism proposed by Wright and Tomita (1965) is of some interest in this respect. These authors visualize the calcium as being located within two protein-phospholipid layers in the axon membrane. Removal of calcium from the outer layer by depolarization allows the entry of sodium ions, while removal of the divalent cation from the inner layer by strong depolarization (i.e. sodium entry or strong stimuli) allows the outward flow of potassium to increase and repolarization to occur. Other authors have also advanced hypotheses to account for the effects of calcium ions on the electrical properties of the axon membranes of arthropods. Thus, Narahashi (1963, 1964) invoked the theory of Tobias, Agin and Pawlowski (1962) to explain the role of calcium in lobster axons, while Adelman and Dalton (1960) have developed the concepts of Mullins(1956) to account for the effects of this ion on the axonal membrane of this crustacean. It is also conceivable that the speculations of Shanes *et al.* (1959) and Judah and Ahmed (1964) may be relevant to these considerations in arthropod axons. It is at the moment impossible to decide how far the above hypotheses are adequate to account quantitatively for the effects of calcium on membrane and action potentials in arthropod axons. A possible structural basis for the action of calcium on the axon membrane is discussed in a subsequent chapter (p. 92) dealing with the protein and lipid organization of arthropod nerve.

The effects of magnesium on resting and action potentials

The effects of variations in the external concentration of magnesium are of interest chiefly for the interactions which this cation undergoes with calcium in influencing the membrane potentials of arthropod axons. In peripheral nerves of both crayfish and lobster, magnesium was found to act synergistically, with calcium, for the axons became spontaneously active only when both these divalent cations were removed from the external solution (Gordon and Welsh, 1948; Adelman, 1956). The membrane potentials of lobster central giant axons, on the other hand, are reported to be unaffected by variations in the magnesium level of the external solution, even in the absence of calcium ions (Dalton, 1958). Magnesium partially substitutes for calcium in the maintenance of the action potential in crayfish giant axons, but has little effect on the resting potential, which is, nevertheless, influenced by the external calcium concentration (Dalton, 1959). These results suggest a dual role for calcium ions on the axon membrane of this species, one being concerned with action potential production (and affected by the magnesium level), the other with the maintenance of the resting potential (being independent of magnesium concentration).

In the stick insect there is a rapid decline in spike height associated with decreased extracellular concentration of magnesium ions, even when calcium is maintained at a constant level (Treherne, 1965 a). Such an effect could result from the effects of magnesium on the axon membrane, which might play a similar role to that of calcium in crustacean axons. In view of the very high extracellular level of this cation the possibility also exists, however, that it might be more directly involved in the production of the action potential (Treherne, 1965 b). From table 6 it will be seen that involvement of magnesium ions in the influx associated with action potential would result in an equilibrium potential of about 29·6 mV. This would produce only a slight increase in the mean potential as compared with that resulting from the movement of sodium ions alone. The involvement of magnesium ions would, however, contribute appreciably to the total ionic current carried by the action potential. It is relevant in this respect to recall that in the muscle fibres of this insect (Wood, 1957) and of the crab (Fatt and Katz, 1953; Fatt and Ginsborg, 1958) the movement of divalent cations has been thought to be involved in the production of the action potential.

The effects of barium ions on resting and action potentials

Although it is known that barium ions can be substituted for those of sodium in the production of action potentials in crustacean muscle fibres (Fatt and Ginsborg, 1958) and also affect the response of lobster (Werman and Grundfest, 1961) and grasshopper muscle fibres (Werman, McCann and Grundfest, 1961), the only arthropod axons which have been studied with regard to the effects of the external concentration of this ion are those of the cockroach *Periplaneta americana* (Narahashi, 1961).

Unlike crustacean muscle fibres (Fatt and Ginsborg, 1958) the giant axons of the cockroach were found to be unable to produce action potentials when the external sodium was replaced by equivalent concentrations of barium ions. With intermediate concentrations the presence of barium ions produced appreciable changes in the shape of the action potential, due to a slowing down of both the rising and falling phases. It is suggested that the decrease in the rate of rise of the action potential in the presence of barium ions is due to a depression of the sodium-carrying system (Narahashi, 1961). The inactivation of the sodium-carrying system, which normally occurs upon depolarization, did not appear to take place in this preparation, for both the active membrane potential and the rate of rise of the action potential were stabilized by the barium ions against the displacement of the membrane potential. The slowing down of the rate of rise of the action potential can also be correlated with a reduced rate of inactivations which will tend to slow down the rate of fall of the action potential. The fact that the spike height is not significantly altered by the presence of barium ions is explained by the delay of inactivation occurring upon depolarization, together with the decreased rate of rise of potassium conductance which would enable the membrane to be depolarized to the normal limit.

The effects of ions on after-potentials

As in other excitable cells the action potentials of arthropod axons have been found to terminate in a relatively slow phase or after-potential. In cockroach giant axons the spike potential is terminated by a positive phase of a few millivolts which is in turn followed by a negative after-potential (Narahashi and Yamasaki, 1960). In crayfish and lobster giant axons, on the other hand, there is no positive phase immediately after the spike potential, which in these species is directly terminated by the negative after-potential (Wantanabe and Grundfest, 1961; Narahashi, unpublished).

Reduction of the external potassium level in experiments with cockroach giant axons resulted in an increase in both the positive phase and the negative after-potential following repetitive stimulation (Narahashi and Yamasaki, 1960). The negative after-potential showed a more pronounced build-up when the potassium level of the bathing solution was reduced in de-sheathed preparations. In these insect axons, as for those of the squid (Hodgkin and Keynes, 1955; Frankenhaeuser and Hodgkin, 1956), the positive phase of the action potential has been postulated to result from a delayed and sustained increase in potassium conductance, while the negative after-potential is explained by the subsequent accumulation of potassium near the axon membrane (Yamasaki and Narahashi, 1959a; Narahashi, 1963). The increase in potassium concentration at the axon surface immediately after the spike has been estimated to be approximately 2·0 mM/l (Narahashi and Yamasaki, 1960). It has been calculated from this figure that the volume of the aqueous layer at the surface of the axon would be roughly equivalent to a thickness of 180 Å (Smith and Treherne, 1963). This value falls within the range of dimensions of 100–200 Å found for the gap between the axon and glial membranes in electron-micrographs of the cockroach central nervous system (fig. 3) and is equivalent to that contained between the inner border of the Schwann cell and the axolemma in the squid giant axon (Geren and Schmitt, 1954), which has been shown to be sufficient to account for the accumulation of potassium necessary to produce the negative after-potential in this preparation (Frankenhaeuser and Hodgkin, 1956).

The decay of the negative after-potential in cockroach giant axons was found to occur as a simple exponential decline, with an average time constant of 9·2 msec (fig. 24a) (Narahashi and Yamasaki, 1960), as compared with the squid giant axons in which the equivalent time constants were between 30 and 100 msec (Frankenhaeuser and Hodgkin, 1956). This difference

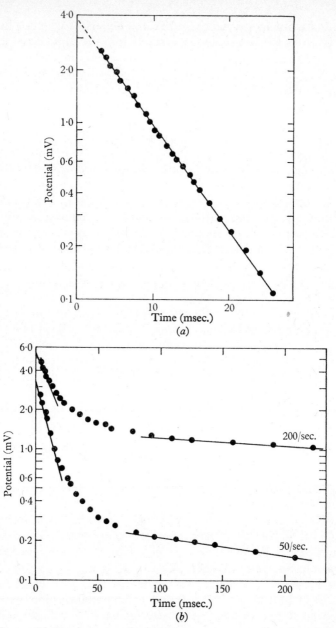

Fig. 24. (a) The time course for the decay of the negative after-potential following a single impulse, recorded from a cockroach giant axon bathed in normal physiological solution. (b) The time course for the decay of the negative after-potential observed in this preparation following a train of impulses at 50 and 200/sec. (After Narahashi and Yamasaki, 1960.)

64

suggests that the diffusion of potassium ions away from the surface of the axon occurs more rapidly in the insect than in the cephalopod axon. It has already been mentioned that one of the unusual features of the concentric extracellular channels surrounding the cockroach axons is the presence of a series of dilatations (fig. 3). It is conceivable that these dilatations could function as reservoirs which, by increasing the volume of the extracellular fluid in the region of the axon surface, could facilitate the dispersal of potassium ions from the axolemma so as to produce a relatively short time constant for the decay of the negative after-potential (Smith and Treherne, 1963). The fact that the decay of the negative after-potential does not assume a simple exponential decline following repetitive stimulation (fig. 24b) (Narahashi and Yamasaki, 1960) could also result from the reservoir-effect of the large extracellular spaces adjacent to the axon surface (fig. 3). It has been suggested that the initial rapid fall in the negative after-potential represents the dispersal of potassium ions from the axon surface, while the subsequent slow decay results from the relatively slow fall in the potassium level as the reservoirs of ions in the large spaces disperse into other parts of the extracellular system (Smith and Treherne, 1963).

The effects of other cations and of metabolic inhibitors on the after-potentials of insect axons have been reviewed in some detail by Narahashi (1963).

CHAPTER 5

Energy Expenditure

THE physiological processes described in the preceding chapters, together with mechanisms of synaptic transmission and the various synthetic reactions of nervous tissues, require the continuous expenditure of energy. This expenditure is reflected in the oxygen consumption of 800·0 mm³/g dry weight/hr (Meyerhof and Schulz, 1929), which is equivalent to 96·0 mm³/g wet weight/hr (Hill, 1930), and a heat production of 0·69 cal/g wet weight/hr (Beresina and Feng, 1933) in the resting peripheral nerve of *Maia squinado*. The value for the oxygen consumption corresponds to a heat liberation of 0·74 cal/g wet weight/hr at the temperature of 20·7° C used by Beresina and Feng, and therefore approximates to the measured heat production in this preparation. Both the oxygen consumption and the heat production of this crustacean nerve represent a level of energy expenditure which is several times greater than that of frog peripheral nerve (cf. Gerard, 1927a, b; Fenn, 1930; Beresina, 1932). This difference may result, in part at least, from the greater proportion of active conducting tissue in the arthropod as compared with the vertebrate nerve (Hill, 1930).

The results of experiments on the oxygen uptake of single crustacean stretch receptor neurones suggest that an appreciable amount of the cell respiration is involved in selective permeability and active transport processes (Giacobini, 1965). In the presence of ouabain and digitoxin, which are known to inhibit sodium transport, it has been shown that there is a marked decline in oxygen consumption. About 60–70 % of the oxygen consumption of the stretch receptor neurone was found to be unaffected by the addition of these metabolic inhibitors. Above this level the oxygen consumption of this preparation can be related to impulse activity.

During activity the heat production of the crustacean nerve increases rapidly, being an order greater than in the unstimu-

66

lated preparation (Beresina and Feng, 1933). Once again this energy expenditure differs from that of amphibian nerve in which the increase in metabolic rate rarely exceeds a factor of 2 following stimulation (cf. Hill, 1932). The relatively small increase of 2·4 times for the oxygen uptake of stimulated crustacean nerve (Meyerhof and Schultz, 1929) probably resulted from fatigue produced by the long periods of stimulation used in these experiments (Feng, 1936). The oxygen consumption during stimulation of crustacean nerve cells has also been shown to be dependent on the ionic composition of the bathing medium (Giacobini, 1965).

The initial phase of heat production in peripheral nerve of *Maia* was found to occur in less than a one second period and to represent only between 1·5 and 2·5 % of the total, the remainder being expended over an extended period of several minutes (Beresina and Feng, 1933). Further analysis with improved techniques showed that a single stimulus resulted in an initial heat production which was synchronous with the impulse (fig. 25). The positive heat production associated with the action potential was estimated to be about $14·0 \times 10^{-6}$ cal/g nerve and the subsequent negative heat to be approximately $-12·0 \times 10^{-6}$ cal/g nerve (Abbot *et al.*, 1958). The total net initial heat measured in previous studies was, in fact, the difference between these very large positive and negative heats.

We have seen in the previous chapters that in arthropod axons, as in those of cephalopods, the rising phase of the action potential probably results from an influx of sodium ions and that the falling phase is associated with an equivalent efflux of potassium from the axoplasm. Now when isotonic solutions of NaCl and KCl are allowed to mix there is a production of heat, which in model experiments with 0·6 M solutions was shown to be 94 cal/mole (Abbott *et al.*, 1958). At 0° C the estimated potassium loss per impulse was approximately $9·0 \times 10^{-8}$ M/g tissue. Taking a figure of 70 cal/M for the heat of ionic exchange, this would give $6·0 \times 10^{-6}$ cal/g/impulse, which is three times the observed net heat and about half the positive heat, indicating that a substantial part of the initial heat during the impulse could be derived from the interchange of sodium and potassium ions. However, more recent experiments have shown that the positive heat does not seem to be increased by the substitution of lithium for sodium ions as the carrier of the

5-2

inward depolarizing current during the action potential (Abbott, Howarth and Ritchie, 1965), despite the fact that the heat of mixing of lithium chloride with potassium chloride is nearly twice that of mixing sodium chloride with potassium chloride (Abbott *et al.*, 1958). It therefore seems unlikely that the heat of ionic mixing is wholly responsible for the initial heat in crustacean nerve, although some effect might be presumed from experiments using barium (which is known to prolong the duration of the action potential; p. 62) and which has been shown to increase greatly the net positive heat production (Abbott *et al.*, 1965).

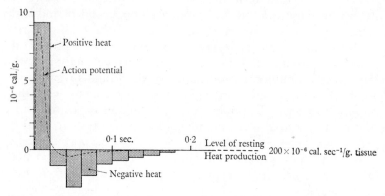

Fig. 25. Heat production in leg nerve of the crab *Maia squinado* at 0° C. The heat produced is represented in blocks of 20 msec intervals, above the resting output of 200×10^{-6} cal sec^{-1}/g tissue. (Redrawn from Rushton, 1962, after Abbott *et al.*, 1958.)

The negative heat associated with the nerve impulse is also difficult to account for. This second phase is unlikely to result from a partial reversal of the ionic interchange, for the replacement of these ions has been shown to be a relatively slow process involving metabolic processes (cf. Hodgkin, 1958), which would, in any case, be expected to result in positive heat production. The possibility cannot be eliminated, however, that the interchanges of some other species of ion may contribute to the negative heat production. It has also been suggested that the negative heat of crustacean nerve may result from endothermic chemical reactions associated with initial anaerobic processes occurring before the onset of the slow oxidative recovery phase (Abbott *et al.*, 1958).

68

The remainder of the additional heat production associated with activity in *Maia* peripheral nerve commences after the passage of the impulse and continues for many minutes. As we have already seen, this phase represents roughly 98 % of the heat produced as a result of nervous activity in crustacean nerve (Beresina and Feng, 1933). This delayed heat production has been identified with the recovery processes involving the secretion of excess sodium ions into the extracellular fluid in exchange for the potassium ions which escaped from the axoplasm during activity (Hodgkin, 1951). The heat produced during prolonged stimulation of *Maia* nerve is roughly equivalent to 55.0×10^{-6} cal/g/impulse (Beresina and Feng, 1933). According to Hodgkin (1951) the energy required to reverse the ionic changes resulting from the passage of one impulse can be represented as:

$$W > CV^2,$$

where C is the capacity of the membrane and V the magnitude of the action potential. With a membrane capacity for crustacean nerve in the region of $1 \mu F/cm^2$ and an action potential of around 120 mV, the energy required to reverse the effect of one impulse could not be less than 1.44×10^{-8} joule/cm^2/impulse. Taking an estimate of 5000 cm^2 for the membrane surface area per gram of nerve in this preparation, the energy required to effect the recovery from the passage of a single impulse becomes 17.0×10^{-6} cal/g. This calculation shows, therefore, that the fibres in *Maia* nerve could maintain a steady state under the conditions of stimulation employed by Beresina and Feng (1933), if it is assumed that an efficiency of about 30 % is required for the extrusion of sodium ions from the axoplasm (Hodgkin, 1951).

Increase in the rate of stimulation of *Maia* produces symptoms of fatigue and a reduction in the rate of heat production (Beresina and Feng, 1933), an effect which might be expected to result from the presence of a rate-limiting chemical reaction involved in the metabolically controlled extrusion of sodium and the associated uptake of potassium ions from the axoplasm. When the nerve was deprived of oxygen, stimulation led to rapid fatigue and decline in heat production. Resting nerve kept in a nitrogen atmosphere, on the other hand, maintained its excitability for much longer periods. The heat production obtained on stimulation of nerves exposed to nitrogen for extended periods was found to be almost indistinguishable from that obtained in experiments carried out in the presence of oxygen (Beresina and Feng, 1933). These results might be expected if the metabolic processes associated with the recovery phase were anaerobic, the immediate source of energy being achieved by oxidative processes.

In insect giant axons the resting potential has been divided into two fractions on the basis of its response to anoxic conditions: the 'aerobic' and the 'anaerobic resting potentials'. Each fraction comprises about half of the total resting potential (Yamasaki and Narahashi, 1957).

In chapter 3 we have seen that in two species of insects, the efflux of sodium from the nerve fibres is reduced in the presence of dilute concentrations of cyanide and 2:4-dinitrophenol (Treherne, 1961e, 1962b, 1965b). As in squid giant axons (cf. Hodgkin, 1958), the sodium extrusion seems to be coupled with an uptake of potassium, for sodium efflux is reduced in the absence of external potassium (Treherne, 1961b) and the influx of potassium is slowed down by the addition of metabolic inhibitors (Treherne, 1965b). The effect of dinitrophenol has been recognized, since the work of Loomis and Lipman (1948), to be that of uncoupling oxidative phosphorylation so as to prevent the formation of adenosinetriphosphate (ATP) and to accelerate its hydrolysis. It is tempting to identify such oxidative phosphorylation as the process which was blocked in the experiments in which arthropod nerves were maintained in the absence of molecular oxygen. Accordingly the almost normal heat production obtained on initial stimulation of resting crab nerve stored in nitrogen (Beresina and Feng, 1933) could be held to result from the anaerobic utilization of stored high-energy phosphate compounds. The reduced heat production and excitability of stimulated crustacean nerve would thus be due to the utilization of these compounds during activity and their non-replacement in the absence of molecular oxygen. The almost immediate effect of dinitrophenol on sodium extrusion would differ from that of the nitrogen atmosphere in that this compound would also cause some hydrolysis of ATP.

There have been no direct experiments in arthropods demonstrating the involvement of high-energy phosphate compounds in sodium extrusion by injecting these substances into the axoplasm, such as have been carried out using squid giant axons

(Caldwell *et al.*, 1960; Baker *et al.*, 1962). There is, however, some more indirect evidence of a coupling between ATP breakdown and ion transport in crab nerve, where it has been found that a particulate preparation of peripheral nerve contains an ATP-ase which requires the presence of both sodium and potassium ions for its activation (Skou, 1957). This enzyme was also found to be inhibited by ouabain, a cardiac glycoside which is known to interfere with ion transport.

Further experiments on the phosphorus metabolism of intact peripheral nerve of *Maia squinado* have also demonstrated the existence of an enzyme system which can utilize energy-rich phosphate compounds (Baker, 1963, 1965). This enzyme system is very similar to the ATP-ase isolated by Skou (1957) from *Carcinus* nerve, being activated by intracellular sodium and extracellular potassium and inhibited by external application of ouabain.

Both cytochrome oxidase and succinic dehydrogenase are concentrated in the mitochondria of lobster peripheral nerve (Foster, 1956). The enzyme mechanisms capable of coupling oxidation and phosphorylation processes are thus available at the axon Schwann-cell surface, where the coupling of energy and impulse conduction occurs.

CHAPTER 6

Carbohydrates

CARBOHYDRATES form the most readily available source of energy in most living systems, and for this reason have been the subject of intensive study in modern biochemistry. In the present chapter this class of compounds will be considered in an attempt to relate carbohydrate metabolism to the specialized organization of the arthropod nervous system. This topic is of especial interest in insects in view of some of the peculiarities in carbohydrate metabolism which are exhibited by this group of arthropods. Carbohydrates also serve as structural elements within the nervous system and this chapter concludes with a short account of the mucopolysaccharides of arthropod nervous tissues.

Blood carbohydrates

There is considerable variation in the reducing values of the haemolymph between different species of arthropod and also fluctuations depending upon the developmental stage and nutritional state of the individual. In crustaceans reducing values equivalent to between 0 and 182 mg/glucose/100 ml have been recorded (cf. Florkin, 1960). The level of reducing sugars in the blood of the insects investigated also shows considerable species and individual variation. In adult honey bees free reducing sugars are present at concentrations of between 1000 and 4000 mg/100 ml (Czarnowski, 1954), although in most species which have been investigated the levels are of the same order as in crustaceans (Wyatt, 1961). Glucose is the characteristic monosaccharide of the insect blood, although fructose has been recorded in the larvae of the bot fly, *Gastrophilus intestinalis* (Levenbook, 1950), and in adult honey bees (Czarnowski, 1954).

In general the greater part of the carbohydrate material in insect blood is present in the form of the non-reducing disaccharide trehalose (Wyatt and Kalf, 1956, 1957; Howden and Kilby, 1956, 1961), which, to choose the extreme example of the larva of the solitary bee *Anthopera*, can be present at concentrations of up to 6554 mg/100 ml (Duchâteau and Florkin, 1959). Trehalose was not detected in the blood of the lobster *Homarus americanus* (Wyatt and Kalf, 1957), although its presence has been demonstrated in whole-body extracts of isopods, amphipods, decapods and arachnids (Fairbairn, 1958).

Sugars

Measurements of reducing substances have yielded extremely high estimates for the carbohydrate content of crustacean peripheral nerves and ganglia (Holmes, 1929). The average 'free carbohydrate' content of *Cancer* peripheral nerve was estimated at 196·1 mg/100 mg tissue as compared with equivalent measurements made on vertebrate nerve which yielded an average value of 42·0 mg/100 mg tissue (Holmes and Gerard, 1929). The estimated level of free sugars in the ventral abdominal ganglia of *Maia* was even higher and averaged 1155·4 mg/100 mg tissue, as compared with a blood concentration of 3–11 mg/100 ml for this species (Berthoumeyroux, 1935). It is possible that these high values may result from a rapid glycolysis occurring in the excised tissues, such as is known to occur in isolated vertebrate cerebral tissues (cf. Geiger, 1962). It would seem desirable to repeat these observations, using modern techniques, to determine whether the apparently high intracellular concentration of free sugars is due to secretory mechanisms or to the metabolism of relatively large indiffusible carbohydrate molecules within the cells of these crustacean nervous tissues.

In *Periplaneta americana*, ^{14}C-labelled glucose and trehalose were identified in the tissues of the abdominal nerve cord following the injection of radioactive glucose molecules into the haemolymph (Treherne, 1960). These measurements were made on washed preparations from which, it is assumed, most of the extracellular labelled material was absent. The estimated figure of 158·0 mg/100 g tissue thus represents only the intracellular content of these sugars, which is, even so, higher than the level of between 86·5 and 108·1 mg/100 g tissue obtained for the reducing substances of mammalian brain tissues (cf. McIlwain, 1959). From the known water content and volume of extracellular fluid of the cockroach nerve cord (p. 14) (Treherne, 1962*b*), it can be calculated that the intracellular

73

level of the exchangeable glucose was around 91·6 mg/100 ml and that for trehalose approximately 183·0 mg/100 ml. The mean concentration of glucose in the blood was 39·5 mg/100 ml, the level of trehalose being 1395·9 mg/100 ml (Treherne, 1960). If the extracellular concentrations of these sugars approximate to those of the haemolymph then it is apparent that the intracellular concentration of glucose is appreciably higher and that of trehalose considerably lower than in the surrounding fluid.

The exchange of sugars between the blood and the tissues of the central nervous system was studied in experiments involving the injection of ^{14}C-labelled glucose into the haemolymph of *Periplaneta* (Treherne, 1960). The injected glucose was rapidly converted to trehalose by the fat body (Treherne, 1958a, b; Candy and Kilby, 1959, 1961) and accumulated in the blood in equilibrium with a low level of glucose. There was initial influx of the high specific activity glucose into the nerve cord immediately after injection, which was followed by equivalent efflux as the specific activity of the monosaccharide in the blood was reduced by the conversion to trehalose. The subsequent slower increase of radioactivity within the nerve cord tissues thus represented the uptake from the blood when the ^{14}C-labelled trehalose was in effective equilibrium with the small amount of glucose in the blood.

The influx of ^{14}C originating from the labelled trehalose and glucose could be brought about by an entry of trehalose or by a relatively rapid movement of molecules from the low level of glucose present in the blood. The data illustrated in fig. 26 show that the measured influx obtained when only glucose was labelled with ^{14}C was equivalent to about 0·08 mM/l nerve cord water/min, which suggests that only about 7 % of the radioactivity in the nerve cord tissues originated from the relatively small glucose pool in the blood. As the trehalose molecules are about seventeen times more concentrated than those of glucose in the blood, this implies that the individual monosaccharide molecules were passing into the nerve cord at about 2·5 times the rate of the disaccharide ones. The free diffusion constant for glucose ($0·68 \times 10^{-5}$ cm^{-2} sec^{-1}) is, however, only 1·2 times that of a disaccharide (e.g. sucrose $0·55 \times 10^{-5}$ cm^2 sec^{-1}). Such an apparent discrepancy between the rates of passage of these two sugars into the nerve cord of this insect could result from the greater restriction to diffusion encountered by the

74

trehalose molecules in the extracellular system (cf. fig. 10) or from any transport of glucose into the cells of this organ. This latter possibility would also serve to explain the apparent concentration gradient which exists for intracellular as compared with external glucose in the central nervous system of this insect.

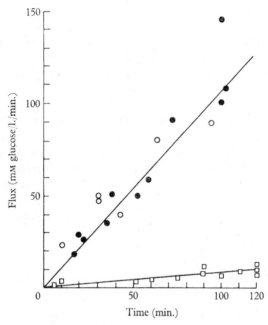

Fig. 26. The calculated influx of radioactivity into the abdominal nerve cord of the cockroach *Periplaneta americana* originating as: (i) trehalose and glucose in the blood of intact insects (closed circles); (ii) trehalose and glucose for connectives isolated in radioactive blood (open circles); (iii) glucose (2·19 mM/l) the trehalose being non-radioactive (open squares). The fluxes are calculated as equivalent mM ^{14}C-labelled glucose/litre nerve cord water/min. (From Treherne, 1960.)

Very little is known of the catabolism of carbohydrates in arthropod nervous tissues. A recent investigation has shown, however, that the enzymes associated with glycolysis and the tricarboxylic acid cycle are very active in the tissues of the cerebral ganglion of the crayfish *Cambarus affinis* (Keller, 1965). The identification of glucose-6-phosphate and phosphoglyceric acid (Heslop and Ray, 1958, 1961), together with the demonstration of glyceraldehyde-3-phosphate dehydrogenase activity

(Boccacci, Natalizi and Bettini, 1960) and the histochemical dehydrogenation of glucose-1-phosphate and α-glycerophosphate (Lambremont, 1962) in insect central nervous tissues, also suggest that carbohydrate metabolism is linked to the tricarboxylic acid cycle (see p. 83) by the conventional glycolytic pathway.

Glycogen

Glycogen represents a major carbohydrate reserve in the tissues of arthropods (cf. Gilmour, 1961; Kilby, 1963; Vonk, 1960). The concentration of this substance in the ganglia of *Maia* has been estimated to average 567·9 mg/100 g tissue and in the peripheral nerve of *Cancer* 973·3 mg/100 g tissue (Holmes, 1929). During anoxia there is an increased hydrolysis of glycogen which is associated with an accumulation of lactic acid in the nervous tissues. More recently the glycogen content of the nervous tissues of the spidercrab *Libinia emarginata* has been found to be as high as 1180·0–1480·0 mg/100 g tissue in ganglia and 1545·0–1680·0 mg/100 g tissue in peripheral nerve (Schallek, 1949), while in the cockroach *Periplaneta americana* the glycogen in the abdominal nerve cord averaged 986·0 mg/100 g tissue (Steele, 1963). These figures represent concentrations which far exceed the values of between 86·5 and 127·9 mg/100 g tissue recorded for vertebrate brain tissues (cf. McIlwain, 1959).

The glycogen level of the abdominal nerve cord of the cockroach has been found to be dependent upon the activity of a hyperglycaemic hormone secreted by the corpora cardiaca (Steele, 1963). Injection of the hormone into the blood caused a rapid fall in the glycogen content of the nerve cord tissues within a period of five hours. This decline, to 14·0 % of the level in control animals, did not appear to result from any effects of the hormone upon the electrical activity of the nerve cord. As in mammals, the degradation of glycogen in insects follows the pathway leading through glucose-1-phosphate to glucose-6-phosphate (Sacktor, 1955; Shigematsu, 1956), and the alterations in glycogenolytic activity appear to result from changes in phosphorylase activity (Steele, 1963).

In *Periplaneta* it has been shown by histochemical techniques that the glycogen is particularly abundant in the perineurium cells of the ganglion and in the cells of the overlying fat body (Wigglesworth, 1960). The glycogen in the perineurium cells appears in electron-micrographs to consist of deeply 'stained' granules (plate 1) (Smith and Treherne, 1963), which fall within the size range (150–400 Å) reported for several vertebrate tissues (Revel, Napolitano and Fawcett, 1960). Smaller deposits were found in the glial processes investing the neurones and axons in the ganglion, and in the circular sheath around the axons in the cercal nerves and abdominal connectives (Wigglesworth, 1960) (fig. 27). Glycogen is also conspicuous in the nerve cells, particularly in the region of the axon cone, where it has the appearance of being associated with the invaginations of the plasma membrane. During starvation the glycogen tends

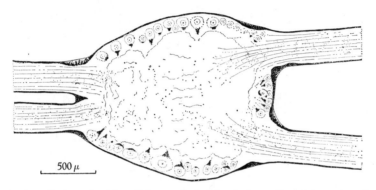

Fig. 27. Horizontal section of the terminal abdominal ganglion of the cockroach *Periplaneta americana*, showing the distribution of glycogen in the perineurium cells of ganglion and nerves, at the inner aspects of the large cell bodies and in the glial components of neuropile and nerve. (From Wigglesworth, 1960.)

to disappear from the tissues of the ganglion, leaving only occasional minute traces in the perineurium cells (fig. 28). Within a few hours of feeding or injecting glucose into the blood the deposits of the polysaccharide were restored. A similar synthesis of glycogen was demonstrated in the central nervous system of mosquito larvae after feeding starved individuals with carbohydrates or amino acids (Wigglesworth, 1942).

It is not clear to what extent the synthesis of glycogen in the perineurium results from the direct metabolism of trehalose or glucose by these cells. In view of the apparent permeability of the neural lamella to ions and molecules, which has already been commented upon (pp. 25 and 43), and the extremely intimate relationship between the fat body and the nerve sheath (plate 1), it is possible that the metabolic processes of these cells may be very closely linked across the fibrous layer. It is conceivable, for example, that the granules seen in the electron-micrographs of the perineurium (plate 1) represent glycogen synthesized in these cells from metabolic intermediaries (some of which may have been degraded from fat-body glycogen)

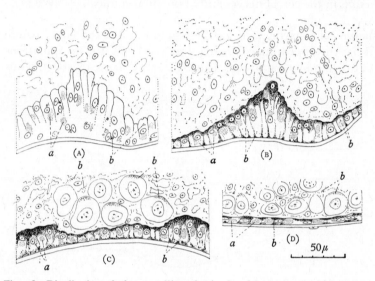

Fig. 28. Distribution of glycogen (*b*) and mitochondria (*a*) in the perineurium cells of the cockroach terminal abdominal ganglion. A, shows the periphery of a ganglion from an insect starved for three weeks. B, the same region 3 hr after feeding a starved individual. C, from the posterior end of the ganglion shown in B. D, from the lateral margin of the ganglion shown in B. (From Wigglesworth, 1960.)

which had diffused across the neural lamella and perineurium plasma membranes from the adjacent fat-body cells. The rapid accumulation of glycogen at the invaginations of the plasma membranes of the cell bodies (Wigglesworth, 1960) suggests these structures may be synthetic sites involved in the rapid transfer of carbohydrates to the neurones of the central nervous system demonstrated with radioactive isotopes (Treherne, 1960).

Mucopolysaccharides

Histochemical tests have revealed the presence of both neutral and acid mucopolysaccharides in the insect central nervous system. The neural lamella of *Rhodnius* (Wigglesworth, 1956) and of the larva of *Galleria* (Ashhurst and Richards, 1964) appears to be composed of collagenous protein embedded in a matrix of neutral mucopolysaccharide. Considerable amounts of acid mucopolysaccharide occur in the extracellular spaces (figs. 1 and 3) of cockroach ganglia (Ashhurst, 1961 *b*), and small amounts are also found in the neural lamella of adult *Galleria* (Ashhurst and Richards, 1964). Electrophoretic evidence suggests that the unit of acid mucopolysaccharide of cockroach ganglia is hyaluronic acid rather than chondroitin sulphate, an identification which is supported by the observed release of glucosamine from this material following hydrolysis (Ashhurst and Patel, 1963). The possible physiological significance of this substance in the extracellular spaces of cockroach ganglia has already been considered (p. 27).

CHAPTER 7

Amino Acids and Proteins

AMINO acids have a multiple role in the arthropod nervous system for, in addition to being incorporated as proteins, these compounds serve as an energy source, and as synaptic transmitter substances, and contribute to the maintenance of the osmotic equilibrium of nerve cells. Some amino acids also have an important role in the metabolism of nervous tissues in forming parts of the pathways involved in the formation of the carbon skeletons of many cell constituents. This chapter gives an account of what is currently known of the distribution and metabolism of amino acids in the peripheral and central nervous tissues of arthropods, and concludes with a brief description of our very limited knowledge of protein metabolism in this animal group.

Blood amino acids

Although there is considerable individual and species variation within the Arthropoda it is possible to distinguish two general types of blood with respect to amino acid composition. In the Crustacea the amino acid content is relatively low and accounts for only a small proportion of the osmotic pressure of the blood (Camien *et al.*, 1951). The total concentration of free amino acids in the blood of the insects which have been investigated tends, on the other hand, to be rather high, reaching in extreme cases levels of up to 2 g/100 ml blood (cf. Duchâteau and Florkin, 1958; Wyatt, 1961; Gilmour, 1961). In some species of Lepidoptera and Hymenoptera the free amino acids may comprise as much as 40–50 % of the osmotic concentration of the blood (Sutcliffe, 1963).

Concentrations of amino acids and related compounds in nervous tissues

There are striking differences between the amino acid compositions of the nervous tissues of the arthropods which have been so far examined (table 10). The gross tissue concentrations of the principal amino acids in the nerve cord of *Periplaneta* are relatively low and, with the exception of proline, are essentially similar to those of vertebrate central nervous tissues. These values are in marked contrast to those for crustacean peripheral nerve, in which the major amino acids are present at much higher orders of concentration. The low level of amino acids in the cockroach nerve cord is not apparently a general characteristic of insect nervous tissues, for the honey bee occupies an intermediate position, in which glutamic acid, alanine and taurine fall within the concentration range of the two crustacean species, the level of aspartic acid being relatively low as in the cockroach.

TABLE 10. *The amino acid content of some arthropod nervous tissues compared with data for the rat. Concentrations expressed as* mM/kg *tissue (wet weight)*

Amino acid	Homarus vulgaris (peripheral nerve)[1,2]	Carcinus maenas (peripheral nerve)[1]	Cancer borealis (peripheral nerve)[2]	Apis mellifica (brain)[3,4]	Periplaneta americana (nerve cord)[5]	Rat (brain)[1]
Aspartic acid	112·0	130·0	—	8·3	3·5	1·7–2·8
Glutamic acid	25·0	35·0	—	50·0	4·5	9·1–12·8
Alanine	33·0	33·0	—	34·6	2·3	0·25–0·96
Taurine	c. 12·0	65·0	—	23·0	—	—
Glycine	35·0	5·0	—	3·5	—	—
Proline	—	—	—	26·1	8·9	0·12–0·15
γ-aminobutyric acid	0·78*	—	0·012–0·016	10·6	2·5	2·0–6·1

* ganglia.

Data from: 1, Lewis (1952); 2, Kravitz, Potter and Gelder (1962), calculated from water content of 86·5% (table 1); 3, Carta, Frontali and Vivaldi (1961); 4, Frontali (1964); 5, Ray (1964); 6, Tallan (1962).

Examination of acid hydrolysates of tissue extracts demonstrated the presence of appreciable amounts of conjugated amino acids among the eighteen detected in the central nervous system of the honey bee (Carta, Frontali and Vivaldi, 1961). Approximately three-quarters of the glutamic acid present and smaller amounts of threonine, serine, glycine, alanine, histidine and arginine were found to be present in a bound form, indicating the presence of appreciable amounts of peptides in this tissue.

The greater part of the massive concentrations of amino acids found in crustacean nerve appears to be situated in the intra-cellular fraction (table 11). There is thus an exceedingly steep concentration gradient between the blood and the cells of lobster peripheral nerve, both for aspartic acid and for the other amino acids studied. A concentration gradient also seems to exist between the cells and the surrounding medium in the

TABLE 11. *The intracellular amino acid concentrations of lobster peripheral nerve and cockroach nerve cord, relative to the blood levels in these species* (mM/kg)

| | *Homarus vulgaris* (peripheral nerve) | | *Periplaneta americana* (abdominal nerve cord) | |
	Blood[1]	Calculated intracellular concentration[2]	Blood[3]	Calculated intracellular concentration[4]
Amino acids				
Aspartic acid	0·17	179·7	0·36	5·98
Glutamic acid	2·02	39·3	0·74	7·63
Glutamine	—	—	3·80	3·99
Alanine	1·14	52·5	1·66	3·54
Glycine	0·80	55·8	9·08	—
Proline	0·29	—	4·40	14·27

Data from: 1, Camien, *et al.* (1951); 2, calculated from Lewis (1952) assuming water content of 865·0 g/kg and extracellular space of 242·0 g/kg tissue; 3, Corrigan and Kearns (1963); 4, calculated from Ray (1964) assuming 733·0 g/kg water content and 21·61 extracellular water (Treherne, 1962 *b*)

central nervous system of the insect species studied, despite the lower amino acid content of these tissues and the relatively high concentration of these compounds in the blood. The actual gradients across the cell membrane for aspartate and glutamate are also likely to be steeper than comparison with the blood in table 11 would indicate, for it is likely that the concentrations of these anions in the extracellular fluid are reduced by the presence of the Donnan equilibrium with the haemolymph (p. 26). The efflux of amino acids from *Carcinus* nerve takes place rapidly, with a half-time of about 5·0 min (Lewis, 1952) as compared to that of 3·0 min for the escape of the extra-cellular sodium into the external medium (Keynes and Lewis,

1951). As the total free amino acids form about 25 % of the dry weight of the nerve it seems unlikely, in view of the rapid exchange with the external medium, that the high intracellular concentrations result from the metabolism of absorbed glucose or other metabolic intermediaries within the cell; they probably result from an active uptake of the amino acids from the extracellular fluid.

The amino acid content of arthropod nerve fibres may show variations which can be related to functional differences, thus in crustacean peripheral nerve the γ-aminobutyric acid concentration has been found to be greatly in excess of that of the excitatory fibres (Kravitz, Kuffler and Potter, 1963). The role of this amino acid in central and peripheral synaptic mechanisms will be considered in a later section (p. 114).

Metabolism of amino acids

The presence of massive concentrations of alanine, aspartate and glutamate in crustacean nerve suggests that these amino acids play an important part in the metabolic processes, owing to their relation with aerobic glycolysis. It also seems obvious that a major function of the dicarboxylic amino acids must be to balance the high internal potassium concentration in these axons, the remaining amino compounds contributing to the maintenance of the high osmotic concentration of the axoplasm (Lewis, 1952). The low amino acid concentration of cockroach nerve (table 11) would accord with this hypothesis, for in this species the intracellular potassium level (table 4) is considerably lower than that for crustacean nerve (table 3), both the level of this cation and the amino acid content approaching that of vertebrate nerve. There is also good evidence of regulation in the concentration and turnover rate of the intracellular amino acids in the nerve cord of *Homarus* in response to changes in the cation composition of the tissues produced by electrical stimulation or treatment with veratrine sulphate (Gilles, 1962; Gilles and Schoffeniels, 1964a).

The rapid appearance of ^{14}C supplied as glucose or trehalose in the carbon skeletons of alanine, glutamate and aspartate in the nerve cords of the cockroach (Treherne, 1960) and the lobster (Gilles and Schoffeniels, 1964a, b) represents circumstantial evidence for the presence of the Krebs tricarboxylic acid cycle enzymes in these tissues. A similar conclusion can

83   6-2

be drawn from experiments demonstrating an accumulation of aspartic and glutamic acids, together with malic and citric acids, in lobster nerves incubated with a solution containing ^{14}C-bicarbonate (Cheng and Waelsch, 1962). The demonstration of carbon dioxide fixation in this arthropod nerve lends support to the suggestion that this compound may serve an 'anabolic' role in nerve metabolism (Monnier, 1952). Thus carbon dioxide fixation might represent a mechanism for the generation of tricarboxylic acid cycle intermediaries and also for the regulation of the rate of turnover of the cycle in arthropod nervous tissues. The tricarboxylic acid cycle enzymes have also been shown to be extremely active in the tissues of the crayfish cerebral ganglion (Keller, 1965).

TABLE 12. *The effect of inhibitors on the free amino acid concentrations in the nerve cord of* Periplaneta americana (*Ray, 1965*). *Concentrations expressed as* mM/kg *tissue* (*wet weight*)

Treatment	Aspartate	Glutamate	Glutamine	Alanine	Proline	γ-amino-n-butyrate
Control	3·5±0·9	4·5±0·25	2·9±0·6	2·3±0·5	8·9±1·7	2·5±0·9
Iodoacetate	2·5	2·0	5·0	2·2	0·9	4·0
Arsenite	2·1	4·3	4·5	12·9	3·9	5·0
Fluoroacetate	3·5	3·0	4·0	8·0	3·3	4·2

The blocking of glycolysis with iodoacetate, which is known to penetrate insect nerve and inhibit glyceraldehyde-3-phosphate dehydrogenase (Boccacci *et al.*, 1960), causes little change in the alanine level, but a marked fall in the proline concentration of the cockroach nerve cord tissues (table 12) (Ray, 1965). It is suggested that the proline is oxidized to provide energy in place of pyruvate and thus help to maintain the level of alanine, which is closely linked to pyruvate.

Blocking of the oxidative decarboxylations of the Krebs cycle with arsenite at pyruvate decarboxylase and α-oxoglutarate dehydrogenase causes a rise in the alanine concentration, which would be expected in the event of the accumulation of pyruvate, and a depletion of proline (table 12). The decline of the latter amino acid was unexpected in view of the probable isolation from the Krebs cycle by the inhibition of the α-oxoglutarate dehydrogenase, and suggests the existence of some other pathway than that through glutamate to α-oxoglutarate. It is conceivable that rise in the level of γ-amino butyric acid in these arsenite, poisoned preparations might occur at the expense of proline, perhaps owing to the activity of glutamic acid decarboxylase, which has been shown to be relatively active in central nervous tissues of the honey bee (Frontali, 1961).

Application of fluoroacetate to cockroach nerve cords causes an increase in the alanine level of the tissues, indicating a failure in the utilization of pyruvate caused by the inhibition of the

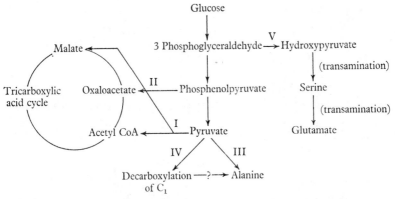

Fig. 29. Metabolic pathways involved in the synthesis of amino acids in the tissues of the ventral nerve cord of the lobster *Homarus vulgaris*. (After Gilles and Schoffeniels, 1964a.)

iso-citric dehydrogenase (Ray, 1965). The actual citric acid level of the nerve cord tissues was found to be from 0·28 μM/g wet weight in normal preparations, as compared with a concentration of 3·0 μM/g wet weight in the fluoroacetate poisoned, ones (Ray, personal communication).

Recent experiments on the ventral nerve cord of the lobster suggest the existence of several routes leading to the synthesis of amino acids from ^{14}C-labelled glucose and pyruvate (Gilles and Schoffeniels, 1964a). Pyruvate seems to follow four different metabolic pathways in these tissues (fig. 29). In addition to an entry into the tricarboxylic acid cycle as activated acetate, it appears that an appreciable amount of pyruvate enters the cycle either directly or through phosphoenolpyruvate. The synthesis of alanine, which occurs as in the insect nervous tissues, is achieved either by a direct pathway from pyruvate or, since there is a difference in labelling of the amino acid according to the localization of ^{14}C in the pyruvate, by a postulated indirect route involving the decarboxylation of C_1 of pyruvate. It also appears that there is an additional pathway for the metabolism of glucose by way of a shunt leading directly to the synthesis of dicarboxylic acids without passing through the Krebs cycle. It is suggested that the hydroxypyruvate shunt might be the possible means by which this synthesis is achieved.

As in vertebrate cerebral tissues, the γ-aminobutyric acid content of arthropod nervous tissues appears to be maintained by decarboxylation of glutamic acid. Glutamic acid decarboxylase activity has been demonstrated in both central and peripheral nervous tissues of the lobster (Kravitz, 1962; Kravitz et al., 1963; Dudel et al., 1963; Frontali, 1964) and in bee brain homogenates (Frontali, 1964). The latter species does, in fact, provide one of the most active tissues so far investigated, the glutamic acid decarboxylase activity being about 2·5 times greater than in mouse brain homogenates.

Protein metabolism

Very little appears to be known of the metabolism of proteins in the nervous system of arthropods. It has been observed, however, that electrical stimulation of peripheral nerve from the crab *Carcinus maenas* causes a decrease in its fluorescence (Ungar and Romano, 1962). The peaks of the emission and activation spectra correspond to those attributed to proteins, an identification which is supported by the effects of denaturing agents, which cause a decrease in fluorescence. It is suggested, therefore, that the decrease in fluorescence associated with the excited state is caused by a change in the configuration of the axon proteins. The reduction in fluorescence is related to the number of stimuli received by the preparation, and returns to the normal level during restoration. It is not known at the moment how these changes in protein configuration are related to the energy transfer processes taking place during excitation.

From the above results it might be expected that electrical activity would be associated with changes in protein synthesis. Stimulation of the isolated crustacean stretch receptor does not, however, affect its total ribonucleic acid (RNA) content, although alterations in the base composition were observed (Grampp and Edström, 1963). Thus both the adenine/uracil and the purine/pyrimidine ratios were significantly increased as a result of stimulation. These increases must have resulted from changes in the synthesis and degradation of the RNA molecule, although the fact that the total RNA content remained constant suggests that these two processes must occur at approximately the same rate in this neurone.

CHAPTER 8

Lipids

LIPIDS are a major constituent of the arthropod nervous system, forming between 13 and 39 % of the dry matter of these tissues. The total lipid content of the crustacean peripheral and central nervous tissues which have been examined tends to be lower than that for the brain of the honey bee, in which the lipid content approaches the figures obtained for several vertebrate species (table 13).

A conspicuous difference between the lipids of arthropod nervous systems and those of vertebrates appears to be the very low level of cerebrosides in the invertebrate tissues. Patterson, Dumm and Richards (1945) have pointed out that the lipids of the honey bee nervous tissues are essentially similar to those of the vertebrate nervous system before the onset of myelination, which suggests that the low level of cerebroside results from the smaller amount of these membrane structures in the arthropod nerve.

Polarized light studies have demonstrated the presence of appreciable quantities of lipid in the axon sheaths in arthropods (Bear and Schmitt, 1937; Richards, 1944), although it is not known what proportion of the total is associated with these structures. Lipids are certainly not confined entirely to the axon sheaths and cell membranes, for these substances have been demonstrated in the cytoplasm of perineurium, and glial and ganglion cells in the central nervous system of *Rhodnius* and *Periplaneta* (Wigglesworth, 1959, 1960). In both these insects there is a clear association of some lipid material with the glial invaginations of the ganglion cells (fig. 30), which, it has been suggested, are concerned with the transfer of lipids to the neurones.

As in axons of other animal groups, the boundary between the axoplasm and the sheath in crustacean nerve appears in electron-micrographs to consist of a double osmophilic line at

TABLE 13. *Lipid content of arthropod nerves compared with those for squid and vertebrate nervous tissues* (mg/100 mg fresh tissue)

| | Limulus[1] | | Libinia[1] | Lobster | | | Apis[4] | Peri-planeta[5] | Loligo[1] | | Rabbit[6] |
	Circum-oeso-phageal ganglion	Nerve cord	nerve cord	Nerve[2] cord	Claw[2] nerve	Claw[3] nerve	brain	nerve cord	Circum-oeso-phageal ganglion	Pallial nerve	whole brain
Total lipid	—	—	—	2·57	1·90	2·18	9·24	—	—	—	10·2–12·4
Phospholipids	1·36	0·99	0·75	1·25	1·01	1·67	4·03	2·0	3·21	0·98	5·0–5·7
Cephalin	0·22	0·18	0·15	0·14	0·12	1·00	2·84	—	1·64	0·28	2·8
Lecithin	0·61	0·42	0·34	1·03	0·86	0·67	0·47	—	1·17	0·42	1·4
Sphingomyelin	0·51	0·45	0·26	0·08	—	—	0·35	—	0·40	0·28	0·9
Choline phospholipids	—	—	—	1·11	0·82	—	—	—	—	—	—
Cerebrosides	0·07	0·09	0·64	—	0·11	0·21	0	—	0·03	0·03	2·4–2·9
Cholesterol	0·28	0·21	0·16	0·26	0·26	0·20	0·31	—	0·72	0·30	2·2
Cholesterol esters	0·02	0·01	0·02	neg.	neg.	—	—	—	0·01	0·01	—
Neutral fat	—	—	—	neg.	neg.	neg.	0·59?	—	—	—	—

Data from: 1, McColl and Rossiter (1950); 2, Brante (1949); 3, Grave (1941); 4, Patterson, Dumm and Richards (1945) assuming water content of 76·3 % (table 1); 5, Heslop and Ray (1961); 6, Spencer (1956).

the axoplasmic surface (Tobias, 1958). Treatment of these axons with proteases and lipases has shown that the structural integrity of the axonal phospholipids is essential for the maintenance of excitability and impulse conduction, although a certain degree of disruption of the surface protein can be tolerated. Both phospholipases and digitoxin were found to produce a marked fall in the electrical constants of these axons (Tobias, 1960). Now in general arthropod nerve appears to differ from that of vertebrates in the possession of an excess of

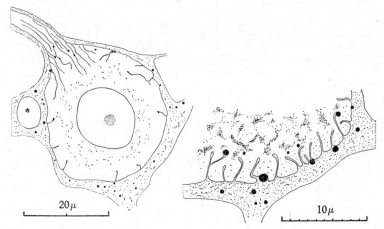

Fig. 30. Distribution of lipid material in cell bodies from the terminal abdominal ganglion of the cockroach *Periplaneta americana*, showing the association of fat droplets with the invaginations of the plasma membrane. This section stained with osmium tetroxide and ethyl gallate. (From Wigglesworth, 1960.)

lecithin as compared with cephalin and in the low level of cholesterol relative to that of lecithin and cephalin (table 13). If it is assumed that there is about five times more lecithin than cephalin, approximately half as much cholesterol as the two phospholipids and about 20 % more cation equivalents available from the calcium present (Salach, 1957) to couple the lecithin molecules, it is possible to postulate the hypothetical structural complex illustrated in fig. 31 for lobster nerve (Tobias, 1958). Such a structure is adequate to explain the results of experiments on the action of enzymes on the axon surface. Accordingly the proteases would be expected to cleave the protein at sites IV, V and VI, the membranes being nevertheless held tenuously in position by the hydrophobic character of the attached lipid layer and the binding between it and the under-lying proteins. Under these circumstances the excitability of the membrane might be expected to be maintained, although it is not clear from this hypothesis how the proteases actually penetrate the lipid barrier.

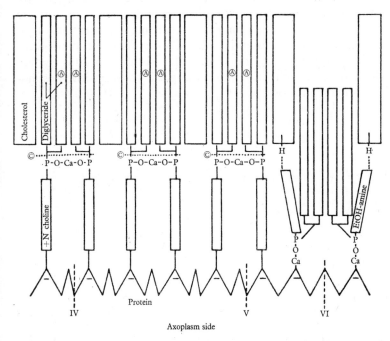

Fig. 31. Hypothetical structure of surface of lobster axon, showing the suggested distribution of lipoidal material. (After Tobias, 1958.)

The action of phospholipase A in inactivating the axon is explained by the cleavage of fatty acid A from its glycerol, while that of phospholipase C can be accounted for in this hypothetical structure by the splitting of the entire diglyceride portion of the molecule from its phosphoryl choline moiety at the sites labelled C in fig. 30.

The above results obtained using lobster axons indicate that the maintenance of normal electrical activity is probably more dependent upon the precise organization of the lipid compo-nents of the membrane than on the organization of the protein elements, which appear to depend to a large degree upon the integrity of the lipid components for their spatial arrangement. It is tempting to speculate from these considerations that the ionic movements taking place in active nerve probably result from changes induced in the lipid elements. The most likely site in Tobias's model to respond to changes in ionic current would seem to be at the protein/lipid interface, where are situated the positive choline-nitrogen link to protein and the negative phosphate-oxygen links to calcium. Experiments with a phospholipid-cholesterol model (Tobias, Agin and Pawlowski, 1962) indicate that the electrical resistance of this system depends upon a combination of inorganic cations with phospholipid acidic groups.

CHAPTER 9

Acetylcholine

THE remaining pages of this monograph are concerned with the role of chemical mediators in the transmission of impulses between nerve cells and at the neuromuscular junction. A brief account has already been given in chapter 1 of the structure and organization of the regions of synaptic contact at the myoneural junction and in the central nervous system of arthropods. The available information indicates that, with the exception of the crayfish median-to-motor giant synapse, the regions of synaptic contact in the arthropod nervous system are essentially similar in many respects to those of vertebrate animals. Despite these apparent similarities there are many points of contrast in the physiological organization of synapses in the arthropods and vertebrate animals which have been studied. In particular there has been considerable controversy as to the precise role of acetylcholine in arthropod nervous tissues. In the present chapter we shall examine the various aspects of the metabolism of acetylcholine and then consider the evidence concerning its role in mechanisms of chemical transmission in the arthropod nervous system.

The distribution of acetylcholine in nervous tissues

The central nervous tissues of the arthropod species so far investigated have been shown to contain concentrations of acetylcholine which are greatly in excess of those found in the central nervous systems of vertebrate animals. In particular the exceptionally high values obtained for the brain and nerve cord of some insect species (table 14) are approximately two orders higher than those found in mammalian central nervous tissues (cf. Long, 1961).

Unfortunately the measurement of acetylcholine activity in several arthropod species has been based on bioassay methods alone, so that the authenticity of the quoted acetylcholine levels remains in doubt in many cases. The possible errors involved in these measurements have been discussed by Chadwick (1963), Colhoun (1963 b), Florey (1961 b, 1962) and Welsh (1961). The identification of acetylcholine in insect nervous tissues has, however, been carried out using various chemical techniques, and there is no doubt of the authenticity of the extremely high concentrations obtained (Augustinsson and Grahn, 1954; Chang and Kearns, 1955; Chefurka and Smallman, 1956; Lewis and Smallman, 1955; Colhoun, 1958 b).

The available evidence indicates that the acetylcholine content of ganglia generally exceeds that of peripheral nerves and connectives. In the crustacean *Cambarus limosus*, for example, the acetylcholine content of the ventral ganglia has been estimated to average 28·0 μg/g tissue as compared with a level of around 8·0 μg/g tissue for the leg nerves and central commissures (Smith, 1939). In the insect *Periplaneta americana* the acetylcholine concentration in the brain tissues exceeded that of the ventral nerve cord (table 14), whilst the thoracic ganglia contained more than three times as much as the thoracic connectives (Colhoun, 1958 b).

The acetylcholine level of arthropod nervous tissues may also show seasonal variations, which depend upon the metabolic state or developmental stage of the animal. In the crustacean *Cambarus limosus* the acetylcholine content of ganglia taken from animals in the Autumn averages about three times that obtained in the Spring (Smith, 1939). In the cecropia silkworm the cholinergic activity of the brain increases during diapause, although the level in the nerve cord ganglia remains at a uniform level (Van der Kloot, 1955).

Studies on crustacean peripheral nerve have indicated that functionally different nerve fibres may contain widely differing concentrations of acetylcholine (Florey and Biederman, 1960). Whole nerves from the chiliped of *Cancer magister* were found to contain 1·7 to 6·7 μg acetylcholine/g tissue. No acetylcholine could, however, be detected in extracts of motor or inhibitory fibres, the total acetylcholine activity of the whole nerve being accounted for by the sensory fibres alone.

The acetylcholine of arthropod nervous tissues appears to exist in two forms. One fraction, commonly known as 'free acetylcholine', can be released into solution by mechanical disruption in the presence of eserine. A second fraction, usually

TABLE 14. *The acetylcholine, acetylcholinesterase and choline acetylase activity of some arthropod nervous tissues*

Species	Preparation	ACh (μg/g tissue wet weight)	AChE (mg ACh hydrolysed/ g/tissue/hr)	ChA (mg ACh synthesized /g/tissue/hr)
CRUSTACEA				
Homarus americanus	Ventral nerve cord	90[9]	7·19–21·87[1]	—
	Leg nerve	46[9]	—	—
Homarus vulgaris	Ventral nerve cord	—	15–20[10]	—
	Ventral ganglia	—	23[11]	—
Astacus fluviatilis	Leg nerve	—	10·0[11]	—
	Brain	18–43[7]	—	—
	Ventral nerve cord	18–39[7]	9–54[7]	—
Cancer irroratus	Ventral ganglia	3·1[12]	—	0·085–0·20[6]
ARACHNIDA				
Limulus polyphemus	Central ganglia	7·5–13·2[12]	—	—
	Leg nerves	10[8]	—	—
	Cardiac ganglion	0·3[11]	—	—
	'Ventral nerve'	—	1·74[13]	—
	'Cardiac nerve'	—	3·61[13]	—
Heteropoda regia	Ganglia	35–50[5]	—	—
INSECTA				
Periplaneta americana	Brain	143[2]	137[4]	50[3]
	Ventral nerve cord	63[2]	270[4]	10[3]
	Leg nerve	1·2 per 60[2]	176[4]	2[3]
Carausius morosus	Nerve cord	100–200[5]	—	—
Musca domestica	Brain	80[3]	—	—

References: 1, Bullock *et al.* (1947); 2, Colhoun (1958b); 3, Colhoun (1958d); 4, Colhoun (1959); 5, Corteggiani and Serfaty (1939); 6, Easton (1950); 7, Florey (1951); 8, Florey (1962); 9, Keyl, Michaelson and Whittaker (1957); 10, Marnay and Nachmansohn (1937); 11, Nachmansohn (1938); 12, Schallek (1945); 13, Smith and Glick (1939).

described as 'bound acetylcholine', requires more drastic treatment with acid conditions or exposure to high temperature to cause release. In the crustacean and arachnid nerve cords examined it has been found that only between 42·6 and 67·1 % of the acetylcholine can be removed by mechanical disruption of the tissues (Schallek, 1945). In insect central nervous tissues, on the other hand, it has been reported that most of the acetylcholine can be liberated by mechanical or osmotic damage (Corteggiani and Serfaty, 1939; Colhoun, 1958a), although in the case of the cockroach thoracic nerve cord it has been claimed that about a quarter of the acetylcholine is present in the 'bound' form (Tobias, Kollros and Savit, 1946).

The concentration of 'free', or readily extractable, acetylcholine in crustacean central nervous tissues has been shown to increase in proportion to the potassium level of the bathing medium (Schallek, 1945). The effect of acetylcholinesterase was eliminated in these experiments by the addition of 10^{-3} M eserine to the experimental solution. The total acetylcholine content of tissues remained approximately constant at the different potassium levels, indicating that the increase in the 'free' acetylcholine occurred at the expense of the 'bound' fraction rather than from any effects of the cation on acetylcholine synthesis, such as those which might result from the stimulating effects of the potassium ions on the citric acid cycle (cf. Kini and Quastel, 1959). Schallek considers that the potassium ions cause a release of 'bound' acetylcholine from specific proteins by cation exchange rather than by effecting permeability changes. The possibility also exists, however, that this effect could result from the increased competition of the potassium with the acetylcholine ions, which might be maintained in solution in subcellular structures by Donnan equilibria.

The precise identity of the so-called 'free' and 'bound' acetylcholine fractions remains obscure. On the basis of experiments demonstrating the release of acetylcholine from nervous tissues by physical means, Colhoun (1958a) favours the hypothesis that this substance is not maintained in chemical combination in insect nerve cells, but is located within 'structural compartments'. By analogy with studies on the vertebrate nervous system (cf. de Robertis, 1958; Whittaker, 1959; Katz, 1962), it would be possible to identify such 'structural compartments' with the vesicular structures demonstrated in the terminal axoplasm of the insect central nervous tissues (plate 2a) (Hess, 1958; Smith and Treherne, 1963, 1965), although the possibility also exists that some acetylcholine could originate from the cytoplasm of the nerve cells. In the light of the recent electron-microscope studies it is also not clear to what extent the apparent differences in the 'bound' and 'free' acetylcholine fractions between different arthropod species result from differences in fragility of the various subcellular components, as well as from the different experimental procedures employed in the

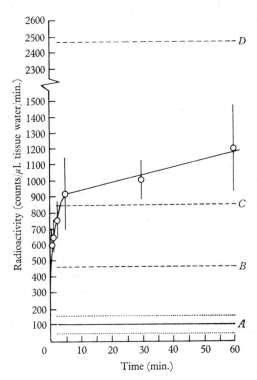

Fig. 32. The uptake of radioactivity per unit volume of tissue for intact cockroach nerve cords exposed to 10^{-4} M ^{14}C-labelled acetylcholine in the presence of 10^{-4} M eserine. *A.* Indicates the level of radioactivity obtained after 1·0 sec exposure to the radioactive solution. The dotted lines indicate the extent of twice the standard error of the mean. *B.* The level of radioactivity which would be expected in the event of the acetylcholine being at the same concentration in the extracellular fluid as in the outside medium. *C.* The level of radioactivity expected in the event of the acetylcholine being distributed in the extracellular fluid according to a Donnan equilibrium with the outside solution. This level was calculated from the concentration ratio observed for inorganic ions in the nerve cord of *Periplaneta* (see p. 27). *D.* The level of radioactivity which would be achieved if the ^{14}C-labelled acetylcholine was at the same concentration in the total tissue water as in the outside medium. The volume of the extracellular water was calculated from the measured inulin space of 18·2 % (see p. 15). The vertical lines drawn through the points represent the extent of twice the standard error of the mean. (From Treherne and Smith, 1965*a*.)

extraction procedures. It would seem that some investigations using sucrose density gradients and ultra-centrifuge fractionation procedures are urgently needed to determine the intracellular distribution of acetylcholine in arthropod nerve cells.

Little is known concerning the diffusion of acetylcholine in arthropod nervous tissues. Experiments with perfused nerve cords showed that with *Homarus* and *Cambarus* 8·3–8·9 % of the total tissue acetylcholine appeared in the perfusate, whereas in the arachnid *Limulus* the acetylcholine in the bathing medium was barely detectable (Shallek, 1945). Nothing is known concerning the differences in the permeability properties to acetylcholine of the crustacean and arachnid nerve sheaths.

On the basis of electrical (Twarog and Roeder, 1956) and toxicological experiments (O'Brien, 1957; O'Brien and Fisher, 1958), it has been suggested that acetylcholine was effectively prevented from penetrating into the central nervous tissues of insects by the presence of a peripheral diffusion barrier, which was identified with the fibrous and cellular nerve sheath (plate 1). More recent experiments using ^{14}C-labelled acetylcholine have shown, however, that this substance penetrates rapidly into the tissues of the intact abdominal nerve cord of the cockroach (Treherne and Smith, 1965a). Uptake in the presence of 10^{-4} M eserine was found to occur as a two-stage process, the initial rapid influx being identified as the penetration into the extracellular system of the nerve cord. The level of radioactivity in the rapidly exchanging fraction was consistent with the hypothesis that the acetylcholine ions were distributed in the extracellular fluid according to a Donnan equilibrium with the haemolymph in these eserinized preparations (fig. 32). In uneserinized nerve cords the extracellular acetylcholine concentration was maintained at a very low level by the activity of the tissue cholinesterases (Treherne and Smith, 1965b).

Acetylcholine synthesis

Synthesis of acetylcholine has been demonstrated in the nervous tissues of three species of Crustacea (Easton, 1950; Walop, 1950, 1951) and in the cockroach *Periplaneta americana* (Tobias *et al.*, 1946; Smallman and Pal, 1957; Colhoun, 1958d, 1959). Synthesis of a cholinergic substance has also been found in the brain of the cecropia silkworm (Van der Kloot, 1955).

Our knowledge of the probable mechanisms involved in the synthesis of acetylcholine largely derives from investigations on whole housefly heads (Lewis, 1953; Smallman, 1956, 1961; Frontali, 1958; Mehrotra, 1961). It is usually assumed with this preparation that the synthetic activity is associated with the cerebral nervous tissues. It appears, as with vertebrates, that the synthesis of acetylcholine occurs as a two-stage process (Smallman, 1956, 1961; Frontali, 1958). In the first stage coenzyme A is converted to acetylcoenzyme A, which in the second reaction contributes the acetyl group for the formation of acetylcholine:

1. $ATP + CoA + acetate \rightleftharpoons acetyl\text{-}CoA + AMP + pyrophosphate$.

2. $Acetyl\ CoA + choline \rightarrow ACh + CoA$.

The second reaction is the limiting process in acetylcholine synthesis and is catalysed by the enzyme choline acetylase.

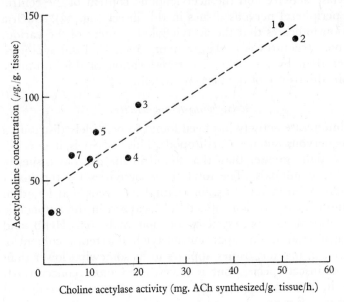

Fig. 33. The relation of choline acetylase activity to acetylcholine concentration in the central nervous tissues of the cockroach *Periplaneta americana*. 1, brain; 2, brain and suboesophageal ganglion; 3, thoracic ganglia; 4, sixth abdominal ganglion; 5, thoracic nerve cord; 6, ventral nerve cord; 7, abdominal nerve cord; 8, thoracic connective. (Plotted from the data of Colhoun, 1958 b, d, 1959.)

The Michaelis-Menten constant (K_m) of the choline acetylase isolated from housefly heads was found to be $4 \cdot 6 \times 10^{-4}$ M for acetyl-CoA (Mehrotra, 1961). This value is markedly different from those obtained for the enzyme extracted from squid ganglia (Reisberg, 1957) and rabbit brain (Smallman, 1958), indicating the existence of variations in the substrate affinity for the enzymes from different sources. Insect choline acetylase

7-2

has also been found to differ from that extracted from squid ganglia (Berman, 1954; Reisberg, 1957) in not being inhibited by iodoacetate (Boccacci *et al.*, 1960). Such differences have been presumed to reflect differences in the protein structure of the enzymes extracted from different sources (Mehrotra, 1961).

There is considerable variation in the choline acetylase activity of tissues from different parts of the cockroach nervous system (Colhoun, 1958*d*, 1959). The activity of this enzyme is highest in brain and suboesophageal ganglion (table 14). There is also fairly close relation between choline acetylase activity and the acetylcholine content of the central and peripheral nervous tissues in this insect (fig. 33). Such a relation indicates that the acetylcholine content of the various nervous structures is probably maintained by local synthesis, rather than by any extensive translocation of this substance within the tissues of the nervous system.

Acetylcholinesterase activity

Cholinesterase activity has been found to be widely distributed in the nervous systems of arthropods. This activity is, in general, substantially greater than that found in the nervous tissues of vertebrate animals. The total tissue activities for the insect species shown in table 14, for example, exceeds that found in the most intense regions of activity localized in the autonomic and central nervous systems of mammals (cf. Hebb and Krnjević, 1962). The apparent acetylcholinesterase content of crustacean nervous tissues appears to be somewhat lower than that of insect species, but is nevertheless more concentrated than in most animal groups which have been investigated (Florey, 1962).

The enzyme responsible for the hydrolysis of acetylcholine in the nerve cord of *Homarus* and ganglia of *Carcinus* is largely the 'specific' cholinesterase which degrades acetylcholine more rapidly than other substrates (Nachmansohn and Rothenburg, 1945; Walop and Boot, 1950). The resemblance of this enzyme to vertebrate acetylcholinesterase is also revealed by its ability to break down acetyl-β-methylcholine, but not benzoylcholine or methyl butyrate (Augustinsson, 1948). Studies on the cholinesterase isolated from whole heads of houseflies show that, although breakdown of butyrylcholine, triacetin and phenylbutyrate occurs, hydrolysis is effected by a single enzyme

(Dauterman, Talens and Asperen, 1962). As with the crustacean nervous tissues there is little evidence for the existence of a 'pseudo' cholinesterase. The acetylcholinesterase found in these housefly preparations can be largely attributed to the central nervous tissues, for it has been found with *Musca* that 91 % of the activity was located in the ganglionic tissue (Stegwee, 1960). The acetylcholinesterase of *Musca domestica* appears to be a very large molecule, with a molecular weight estimated to be in the region of 3–4 million (Dauterman *et al.*, 1962). The turnover rate for this enzyme has been measured to be in the neighbourhood of 110,000.

A recent study using three decapod crustaceans has shown that the nervous tissues possess two electrophoretically distinct enzymes which hydrolyze acetylthiocholine and which have been designated as acetylcholinesterases (Maynard, 1964). The more rapidly migrating form predominates in the central ganglia and connectives, the slower enzyme being the principal one active in peripheral nerves. Nothing is known of the cellular localization of these acetylcholinesterases, although the possibility that the rapidly migrating fraction might be associated with synaptic regions seems to be eliminated by the observation that this form predominates in the interganglionic connectives, where there are no synaptic regions.

Limitations of space prevent further discussion of the great volume of research which has been carried out in recent years on the chemical properties of acetylcholinesterases in arthropods. This work, which has been chiefly concerned with various insect species, has, however, been extensively reviewed by Chadwick (1963).

The distribution of cholinesterases in nervous tissues has been studied by histochemical methods, using the light microscope, in the nervous systems of the lobster *Homarus americanus* (Maynard and Maynard, 1960), and the insects *Rhodnius prolixus* (Wigglesworth, 1958), *Musca domestica* (Molloy, 1961) and *Periplaneta americana* (Iyatomi and Kaneshina, 1958; Winton, Metcalf and Fukuto, 1958; Kaneshina, 1961; Hamori, 1961). These studies demonstrate intense cholinesterase activity in neuropile, together with a distribution in the glial sheaths encapsulating the neurone perikarya and in the perineurium of the nerve sheath. The enzyme appeared to be absent in regions of neuromuscular transmission in the insect species.

The greater resolution obtained with the electron-microscope has revealed the following distribution of eserine-specific esterase activity in the abdominal nerve cord of the cockroach *Periplaneta americana* (Smith and Treherne, 1965):

(1) in the glial sheaths of the axons in the connectives and cercal nerves;

(2) in the glial folds encapsulating the neurone perikarya in the ganglion;

(3) in localized areas along the membranes of axon branches within the neuropile, frequently in association with focal clusters of synaptic vesicles (plate 3).

In addition, activity of eserine-insensitive esterase has been found in glial elements in the terminal abdominal ganglion, connectives and cercal nerves. Such unspecific esterase, which is, however, inhibited by tetraethyl pyrophosphate and tri-ortho-cresyl-phosphate (Colhoun, 1960; Stegwee, 1960), does not appear to be involved in nervous activity (Colhoun, 1960).

Some recent experiments have demonstrated the impressive ability of the above cholinesterase system in the hydrolysis of acetylcholine and the reduction of the extracellular concentration of this substance. As has already been mentioned, the influx of ^{14}C-labelled acetylcholine into the extracellular system of the cockroach nerve cord has been found to occur extremely rapidly, with a half-time in the region of 50 sec (Treherne and Smith, 1965a). Despite this rapid influx the extracellular acetylcholine concentration was found to be reduced to a calculated level of around $8 \cdot 1 \times 10^{-5}$ M in terminal abdominal ganglia bathed with a solution of 10^{-2} M acetylcholine (Treherne and Smith, 1965b). In these experiments the products of the hydrolysis of acetyl-^{3}H-choline accumulated largely as acetate and glutamate within the nerve cord tissues. These results indicate that an appreciable hydrolysis of the applied acetylcholine occurred towards the periphery of the nerve cord. It was shown, for example, that when whole nerve cords were bathed in a solution of 10^{-2} M acetylcholine the concentration of this substance showed only a slow approach to the level obtained after a $1 \cdot 0$ sec exposure to the radioactive solution in the presence of 10^{-4} eserine (fig. 34). A $1 \cdot 0$ sec exposure to the experimental solution is known to produce a level of radioactivity which is approximately equivalent to the amount of labelled acetylcholine contained in the nerve sheath following a prolonged exposure to the labelled solution (Treherne and Smith, 1965 *a*). It appears, therefore, that a considerable hydrolysis of the applied acetylcholine must have occurred

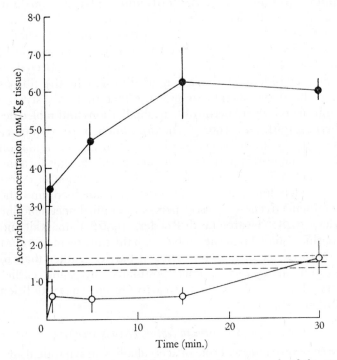

Fig. 34. The uptake of acetylcholine in whole nerve cords of the cockroach *Periplaneta americana* which were exposed to a $1 \cdot 0 \times 10^{-2}$ M solution of radioactive acetylcholine for varying periods. The concentrations of the labelled acetylcholine were taken from chromatograms of the tissue extracts. The continuous horizontal line represents the concentration of radioactive acetylcholine obtained following a $1 \cdot 0$ sec exposure to the experimental solution in the presence of 10^{-4} M eserine. The broken line represents the extent of twice the standard error of the mean. The closed circles represent the acetylcholine concentration in nerve cords exposed to the experimental solution in the presence of 10^{-4} M eserine; the open circles show the accumulation of radioactivity in untreated preparations. The vertical lines drawn through the points indicate the extent of twice the standard error of the mean. (From Treherne and Smith, 1965 *b*.)

within the nerve sheath. The peripheral sites of cholinesterase activity, which are localized on the glial membranes bordering extracellular channels (plate 3) (Smith and Treherne, 1965; Treherne and Smith, 1965 *b*), would seem to be strategically placed to effect hydrolysis of applied acetylcholine in the perineurium and connective tissue layer of the nerve sheath (plate 1). It is claimed that a portion of the cholinesterase of the central nervous tissues of this insect appears to be relatively free and can diffuse from nerve cords into the bathing medium (Mikalonis and Brown, 1941), although this result was not confirmed by Colhoun (1958c). In any event the large cholinesterase molecules, with a molecular weight in the region of 3–4 million (Dauterman et al., 1962), might be expected to encounter considerable restriction to diffusion in the fibrous layer of the nerve sheath, despite the fact that this layer is known to be relatively permeable to smaller ions and molecules (Treherne, 1961a, e; 1962b; 1965b). Such a restriction to diffusion in the fibrous layer would result in an accumulation of this enzyme in this portion of the nerve sheath, thus forming an extremely efficient mechanism for the destruction of extraneous acetylcholine penetrating from the outside medium. The relative ineffectiveness of insect nerves to applied acetylcholine (Roeder, 1948; Twarog and Roeder, 1956; Yamasaki and Narahashi, 1960; Treherne, 1962c) would thus seem to result from the presence of this 'biochemical barrier' rather than to the peripheral *diffusion* barrier postulated by some earlier workers (Twarog and Roeder, 1956; O'Brien, 1957; O'Brien and Fisher, 1958).

The cholinergic system in central nervous synapses

The precise physiological role of acetylcholine in synaptic transmission in the arthropod central nervous system has been a subject of some speculation and controversy. The difficulties involved in postulating the existence of the cholinergic system in synaptic transmission in arthropod ganglia have been emphasized by several authors (Yamasaki and Narahashi, 1960; Welsh, 1961; Hebb and Krnjević, 1962; Florey, 1962).

The criteria by which cholinergic transmission is identified have been listed by Hebb and Krnjević (1962) as follows:

(1) Pharmacological evidence showing that applied acetylcholine mimics the effect of stimulation of the nerve. It should also be shown that the activity produced by both kinds of stimuli can be modified by appropriate drugs.

(2) Demonstration of a release of acetylcholine from nerve endings during activity.

(3) Biochemical and histochemical evidence that acetyl-

LEGENDS TO PLATES 1–3

1. Electron-micrograph showing a longitudinal section of the terminal abdominal ganglion of the cockroach, *Periplaneta americana*. The fibrous portion of the nerve sheath, the neural lamella (*nl*), is closely overlaid by a portion of a fat body cell (*FB*) which contains clusters of glycogen deposits and large cavities (*). The bulk of the field is occupied by cytoplasm of the perineurium and shows: the nucleus of one cell (PN_1) at *n*, the surrounding cytoplasm being filled with mitochondria and glycogen granules. PN_2 represents a portion of another perineurium cell containing numerous mitochondria, but no granular glycogen deposits, which are, however, tightly packed elsewhere in the cell (*g*). A glial cell process (*GL*) is seen, which associates with the axons (*ax*) in the lower right hand corner. Glycogen granules also appear in the glial cytoplasm. × 13,000. (From Smith and Treherne, 1963.)

2 (*a*) Synaptic region from the neuropile of the terminal abdominal ganglion of the cockroach *Periplaneta americana*. The plasma membranes of the two axons (ax_1 and ax_2) are separated by a gap of about 100 Å. The cytoplasm of ax_1 contains numerous synaptic vesicles which are concentrated at intervals against the plasma membrane. These foci of synaptic vesicles are mirrored in the opposite axon (ax_2) by zones of slightly increased density (arrows). × 55,000. (From Smith and Treherne, 1963.)

(*b*) Portion of a motor terminal on a coxal muscle fibre of the bee, *Apis mellifera*. This shows concentrations of synaptic vesicles (*sv*) in the terminal axoplasm. At the regions of synaptic contact (arrows) the plasma membranes of the axon and muscle fibre are separated by a gap of only *c*. 100 Å × 33,000. (From Smith and Treherne, 1963.)

3 (*a*) An electron-micrograph of a field at the periphery of a connective in the abdominal nerve cord of *Periplaneta americana*. This material was incubated to reveal sites of esterase activity and shows: a portion of the fibrous nerve sheath (*Sh.*) with an underlying layer of glial cytoplasm (*gl.*) and numerous interaxonal glial processes. Small dense granules of the reaction product are present in association with the membranes of the glial processes (arrows). These sites of esterase activity are inhibited by eserine. (From Treherne and Smith, 1965*b*.)

(*b*) An electron-micrograph of a field within the neuropile of the terminal abdominal ganglion of *Periplaneta* treated to demonstrate sites of eserine-sensitive esterase activity. It is suggested that these membrane-associated sites of enzymatic activity (1–5), together with the vesicles (*sv*) adjoining them, represent synaptic regions between presynaptic (*pr.*) and postsynaptic (*po.*) elements. Four focal clusters of vesicles (*) are shown in this picture. The presumed postsynaptic member contains mitochondria (*m*), but not synaptic vesicles. × 70,000. (From Smith and Treherne, 1965.)

(*c*) An enlargement of region 3 in plate 3*b* showing that the reaction product appears to be situated on both pre- and postsynaptic membranes; a narrow gap (arrow) separating the two regions of cholinesterase activity. × 115,000.

facing p. 104

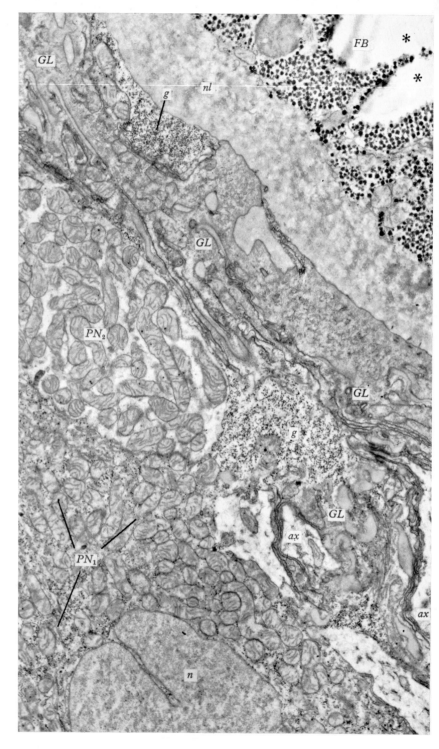

I

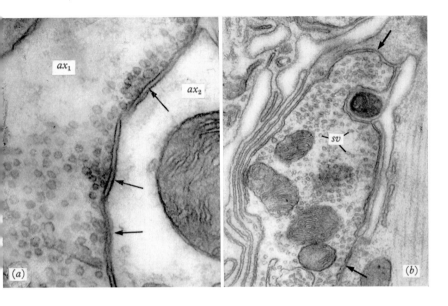

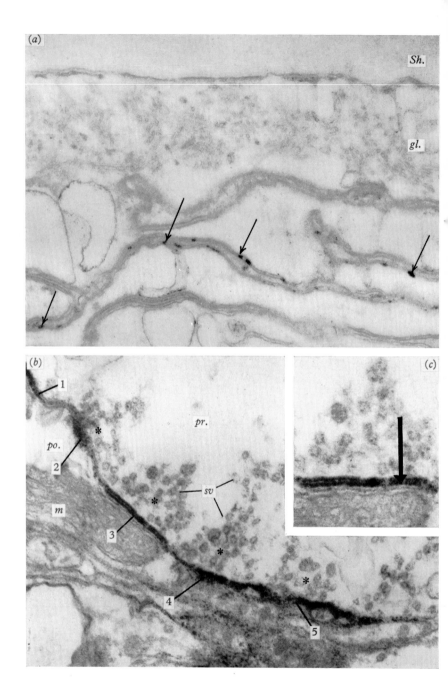

3

cholinesterase and choline acetylase are present in the nerve and that the acetylcholinesterase is located at the receptor sites of the cells.

One of the main difficulties involved in applying the first of these criteria for arthropod central synapses has been the apparent insensitivity to applied acetylcholine of the preparations used. Synaptic transmission in the abdominal nerve cord of crustacean species studied has been found to be remarkably insensitive to the addition of acetylcholine to the bathing medium (Prosser, 1940 b; Schallek and Wiersma, 1948; Turner, Hagins and Moore, 1950). In the terminal abdominal ganglion of the cockroach *Periplaneta americana*, synaptic conduction was found to be unaffected by concentrations of up to 10^{-2} M acetylcholine in the bathing medium (Roeder, 1948). This level of acetylcholine is more than two orders higher than the concentration which evokes the discharge of impulses in vertebrate sympathetic ganglia (cf. Bronk, 1939).

The reported insensitivity of the cockroach terminal abdominal ganglion to applied acetylcholine does not appear to result from an exclusion of this ion from the extracellular fluid by a peripheral diffusion barrier associated with the fibrous and cellular nerve sheath. As has already been mentioned (p. 98), acetylcholine penetrates rapidly into the extracellular system in eserinized preparations (Treherne and Smith, 1965 a). The insensitivity of this preparation to acetylcholine can, however, be attributed to the extremely rapid hydrolysis of this substance by the peripheral cholinesterase system of the ganglion (see p. 102). Ganglia bathed with a solution of 10^{-2} M acetylcholine showed a calculated extracellular concentration of this ion of only about $8\cdot1 \times 10^{-5}$ M (Treherne and Smith, 1965 b).

The failure of acetylcholine to affect central synaptic transmission in the crayfish (Schallek and Wiersma, 1948) may, as Florey and Biederman (1960) point out, result from the fact that only the action on synapses between the giant fibres and motor axons was studied. This type of synapse has been shown to be peculiar in the very close apposition of the pre- and post-synaptic membranes (Robertson, 1953, 1961) (see p. 11), resulting in the formation of an 'electrical' rather than a 'chemical' transmission mechanism (Furshpan and Potter, 1959 a).

The above results indicating an apparent lack of sensitivity to applied acetylcholine do not by themselves, therefore, eliminate the possibility that cholinergic mechanisms are involved in synaptic transmission in the arthropod central nervous system.

The presence of eserine in the bathing medium has been shown to exert an excitatory and blocking effect on crustacean central synapses (Prosser, 1940*b*; Schallek and Wiersma, 1948; Turner *et al.*, 1950). In the terminal abdominal ganglion of the cockroach treatment with eserine and other anticholinesterases caused a marked increase and prolongation of the excitatory post-synaptic potentials, which were superimposed upon a prolonged after-discharge, and eventual synaptic block (Yamasaki and Narahashi, 1960).

In the desheathed cockroach ganglion, acetylcholine becomes effective in discharging impulses at concentrations of around 10^{-4} M, as compared with the level of 10^{-2} M obtained in un-eserinized preparations (Twarog and Roeder, 1957; Yamasaki and Narahashi, 1960). The threshold concentration for this insect preparation is, however, much higher than that observed with vertebrate sympathetic ganglia, in which concentrations of acetylcholine as low as $5 \cdot 0 \times 10^{-7}$ M are effective in discharging impulses in the presence of eserine (Emmelin and MacIntosh, 1952). The results for the crustacean species which have been studied differ from those obtained with the insect terminal abdominal ganglion in that the addition of acetylcholine to an eserinized preparation had no further effect (Prosser, 1940*b*; Schallek and Wiersma, 1948; Turner *et al.*, 1950). The action of drugs, such as nicotine, amines, quarternary ammonium bases and DFP, was, however, found to be essentially similar to that found in vertebrate autonomic ganglia, where the role of acetylcholine in synaptic transmission is well established (Schallek and Wiersma, 1948, 1949; Shallek, Wiersma and Alles, 1948).

The second criterion of cholinergic transmission listed by Hebb and Krnjević (1962), the release of acetylcholine from nerve endings during activity, appears to have been investigated in only one arthropod species. Electrical stimulation of isolated cockroach nerve cords in the presence of eserine was found to increase the acetylcholine content of the tissues (Colhoun, 1958*c*). Stimulation of the anal cerci by air puffs, on the other hand, produced no appreciable difference in the acetylcholine content of normal and stimulated terminal abdominal ganglia. It seems reasonable to assume that the latter effect might result from the extremely minute amount of acetylcholine produced at the synapses involved, relative to the total tissue content of this substance.

The third criterion of cholinergic transmission listed above concerns the demonstration of cholinesterase and choline acetylase activity in the nervous tissues. We have already seen (p. 95) that both of these systems exist in remarkably high concentrations in the nervous tissues of the arthropod species so far investigated. In particular it is significant that eserine-sensitive cholinesterase activity has been demonstrated in the neuropile of cockroach ganglia in localized areas of axon branches in association with the foci of synaptic vesicles (plate 3) (Smith and Treherne, 1965).

On the basis of the rather fragmentary evidence outlined above it is clear that only in the case of the cockroach central nervous system is it possible to postulate the existence of the conventional cholinergic system with any degree of certainty. Even here there are difficulties. The extremely high concentration of acetylcholine necessary to evoke the discharge of impulses in eserinized insect ganglia is, for example, an obvious point of difference from the cholinergic system in vertebrate autonomic ganglia. This effect can, however, be quite adequately accounted for by the postulation of a relatively low sensitivity of the postsynaptic membrane to acetylcholine. Such a degree of insensitivity can perhaps be correlated with the very high acetylcholine and cholinesterase content of the insect central nervous tissues examined.

The results obtained with the crustacean central nervous system are perplexing. The effects obtained with various drugs together with the relatively high concentrations of apparent acetylcholine and cholinesterase demonstrated in these tissues might be taken to indicate the existence of cholinergic transmission mechanisms. The lack of effect of applied acetylcholine in eserinized preparations is, however, difficult to reconcile with the conventional cholinergic system. The low concentration of 'pseudo' cholinesterases in the crustacean central nervous tissues examined makes it difficult to explain this effect by postulating an appreciable hydrolysis of applied acetylcholine by these enzymes. One possibility which remains is that transmission may be mediated by some analogue of acetylcholine which exhibits some of the physiological properties of authentic acetylcholine. It is clear that further investigation, utilizing microelectrode and other techniques used in studies on vertebrates, is required in this field.

The cholinergic system in neuromuscular transmission

In vertebrate muscle fibres the application of curare is known to block neuromuscular transmission by depressing the end-plate potential of the post-synaptic membrane below the threshold required for a propagated response. This effect is believed to result from the specific blocking of the acetylcholine-mediated system. In arthropod species, on the other hand, application of curare, and other drugs known to affect vertebrate neuromuscular transmission, has been found to be without effect (Straub, 1900; Katz, 1936; Ellis, Thienes and Wiersma, 1942; Roeder and Weiant, 1950; Usherwood, 1963). Application of acetylcholine to the neuromuscular junction of the skeletal muscles of crustacea (Bacq, 1935, 1947; Katz, 1936; Ellis *et al.*, 1942); and insects (Harlow, 1958; Usherwood, 1963) has also been found to be ineffective. The report that injection of acetylcholine into crayfish claws causes rapid and transient closure (Florey and Florey, 1954) can be explained by postulating that this substance stimulates sensory cells, the axons of which excite motor axons by ephaptic transmission (Florey and Biederman, 1960).

Acetylcholine has been found to be absent in crustacean motor fibres (Florey and Biederman, 1960), while cholinesterase activity is also exceedingly low in the skeletal muscles of insect and crustacean species (cf. Florey, 1962; Colhoun, 1963 *b*). Histochemical investigation has in addition demonstrated an absence of esterase activity at the nerve terminal of insect motor end-plates (Wigglesworth, 1958; Iyatomi and Kanehisha, 1958; Hamori, 1961).

These results clearly indicate that in arthropod skeletal muscle, at least, neuromuscular transmission is not mediated by the conventional cholinergic system.

Acetylcholine has been demonstrated to have an excitatory effect on crustacean and insect hearts (cf. Maynard, 1961; Welsh, 1961; Davey, 1964). Some of the classical blocking agents for vertebrate autonomic ganglia have also been found to be effective on these preparations, indicating that acetylcholine might be involved in the control of the arthropod heart. However, it has been reported that cardiac accelerator fibres and the cardiac ganglion of the lobster (*Homarus americanus*) contain no detectable amounts of acetylcholine (i.e. less than $0 \cdot 1\ \mu g/g$ tissue) (Florey, 1961 *b*). Furthermore the effects of atropine (which blocks the action on the heart) do not alter the response to stimulation of the cardio-accelerator fibres, while application of eserine is also without effect. It has been concluded on the basis of these results that cholinergic transmission is not normally involved in cardio-acceleration in this crustacean (Florey, 1961 *b*). It should also be noted that the heart of the cladoceran *Daphnia* does not exhibit a normal response to the classical cholinergic drugs (Baylor, 1942; Prosser, 1942; Bekker and Krijgsman, 1951).

CHAPTER 10

Other Possible Transmitter Substances

AN impressive array of pharmacologically active substances has been isolated from the tissues of various arthropod species. In many cases, however, the metabolism of these compounds and their precise physiological role in nervous activity remain obscure. In this chapter attention will be confined to those compounds which, in addition to acetylcholine considered in the preceding chapter, are likely to be involved in synaptic transmission mechanisms in arthropod central and peripheral nervous systems.

Catecholamines

Both adrenalin and nor-adrenalin have been identified in whole body extracts from a variety of insect species (Östlund, 1954; Euler, 1961). These studies have shown that the nor-adrenalin content is greatly in excess of that for adrenalin in the insect tissues examined. In crustacean species, on the other hand, these amines have been found to occur only in very small concentrations. Dopamine, the precursor of nor-adrenalin, has also been shown to occur in high concentrations in the insect body (Östlund, 1954).

The corpus cardiacum of the cockroach *Periplaneta americana* has been demonstrated to contain an adrenalin-like material, which it is claimed differs from this substance in some chemical properties (Cameron, 1953; Gersch, Unger and Fischer, 1957). The concentration of such material in the gland is reduced by electrical stimulation (Barton-Browne *et al.*, 1961).

The possibility of the existence of adrenergic synaptic mechanisms in arthropod central nervous tissues is demonstrated by the work of Twarog and Roeder (1957) on the terminal abdominal ganglion of the cockroach. These workers recorded bursts of asynchronous action potentials of low amplitude following application of adrenalin and nor-adrenalin at concentrations of 10^{-4} M. After-discharge and synaptic blocking were observed at concentrations of between 10^{-3} and 10^{-2} M. The high threshold level necessary to evoke impulses in this preparation may, perhaps, be correlated with the relatively high concentrations of these substances demonstrated in insect tissues (Östlund, 1954). Further evidence for the possible involvement of catecholamines in synaptic transmission in the central nervous system comes from some recent experiments on the effect of 3-hydroxytyramine (dopamine; 3,4-dihydroxyphenylethylamine) on the electrical activity of the cockroach terminal abdominal ganglion (Gahery and Boistel, 1965). In this study the above substance was found to induce bursts of activity, which were propagated along the nerve cord, when the ganglion was irrigated with 5.0×10^{-5} M solution. This substance did not have any effect, however, on synaptic transmission between the cercal nerves and the giant fibres of the nerve cord. The fact that adrenalin and nor-adrenalin exert an effect on the cercal synapse (Twarog and Roeder, 1957) suggests the possible existence of functional differences between catecholamines at different synaptic sites within the ganglion. It should also be noted that recent electron-micrographs have shown that not all of the apparent synaptic regions in the ganglion of this insect are associated with membrane-bound cholinesterase activity (Smith and Treherne, 1965). Such apparently non-cholinergic synapses could be regarded as possible candidates for transmission mechanisms involving catecholamines.

3-Hydroxytyramine has been found to be exceedingly active as an inhibitor of the crustacean stretch receptor neurone, being over a hundred times more effective than GABA (McGeer, McGeer and McLennan, 1961). This inhibitory action has been presumed to result from the configuration of the molecule, which approximates to that of GABA in the possession of both acidic and basic groups (p. 116). The precise locus of action of the 3-hydroxytyramine is uncertain and it is not known whether the greater activity of this catecholamine, as compared with GABA, results from an effect on axonal conduction as well as on the dendrites and cell body or from differences in receptor sites on the same active membrane.

Adrenalin and nor-adrenalin have been found to be ineffective at the lobster neuromuscular at concentrations up to 10^{-4} M (Bergmann, Reuben and Grundfest, 1959), although catecholamines have been found to excite both crustacean (cf. Maynard,

1961) and insect hearts (cf. Davey, 1964; Jones, 1964). With the isolated cockroach heart, for example, the threshold concentration of adrenalin is around 10^{-7} M (Krijgsman and Krijgsman-Berger, 1951; Naidu, 1955). As with vertebrates this effect seems to be a hormone-like one, for the excitatory effect of adrenalin does not appear to be mediated by the nervous system in this insect preparation (Davey, 1964).

5-Hydroxytryptamine

5-Hydroxytryptamine has been identified in nervous tissues of some arthropod species (Florey and Florey, 1954; Welsh and Moorhead, 1960; Gersh et al., 1961; Colhoun, 1963a) (table 15), although the concentrations were considerably lower than those obtained for the nervous tissues of annelids and molluscs (Welsh and Moorhead, 1960). The highest concentrations were found in the arachnid Limulus polyphemus, and it has been tentatively suggested that in arthropods, as in molluscs, less 5-HT is required in the nervous tissues of the more highly evolved groups (Welsh and Moorhead, 1960). Synthesis of 5-HT has been demonstrated in extracts of insect nervous tissues (Colhoun, 1963a) and decarboxylation has been shown to occur in homgenates of ganglia of Cancer borealis (Welsh and Moorhead, 1960). The decarboxylase activity in these crustacean ganglia was, however, lower than that obtained in molluscan nervous tissues.

5-HT has been found to be concentrated in the crustacean pericardial organ (Maynard and Welsh, 1959; Welsh and Moorhead, 1960), together with a compound identified as 5,6-dihyroxytryptamine (Carlisle, 1956; Carlisle and Knowles, 1959), and in the insect corpus cardiacum (Gersh et al., 1961; Colhoun, 1963). It is not known to what extent the 5-HT is associated with axons in these structures, but its function appears to be mainly that of a neurohumor.

There is little concrete evidence that 5-HT is involved in synaptic transmission in the central nervous system of arthropods. This substance has been found to be ineffective at the cercal synapse when applied to the terminal abdominal ganglion of the cockroach (Twarog and Roeder, 1957), although this does not preclude the possibility of its functioning in other synaptic regions.

5-HT and some related compounds have been shown to

112

affect neuromuscular transmission in the fibres of the metathoracic extensor tibialis and flexor tibialis of the locust *Locusta
migratoria* (Hill and Usherwood, 1961). The inhibition which
these substances produce operates at the synaptic membranes,
for it was shown that it did not result from blockage of transmission in nerve fibres or from diminution of the responsiveness of
the electrically excitable component of the muscle fibre membrane. The activity of the tryptamine analogues on neuromuscular transmission in this insect is similar to that observed

TABLE 15. *Concentration of 5-hydroxytryptamine in the nervous tissues
of some arthropod species.* (*Data from Welsh and Moorhead, 1960*)

Class	Species	Tissue	5-HT content (μg/g fresh tissue)
CRUSTACEA	*Orconectes virilis*	Nerve cord	< 0·10
	Homarus americanus	Nerve cord	< 0·02–0·03
		Leg nerves	< 0·03
	Cancer borealis	Ventral ganglia	0·02–0·08
		Leg nerves	< 0·04
		'Brain'	0·08
		Pericardial organs	3·2–4·0
ARACHNIDA	*Limulus polyphemus*	Heart and cardiac ganglion	0·09
		Nerve cord	0·10–0·20
		Leg nerves	0·15–0·29
INSECTS	*Blaberus gigantea*	Nerve cords	< 0·02

with the *Venus* heart preparation (Greenberg, 1960), in which
it was shown that the simplest structural requirement for 5-HT
activity was a flat aromatic nucleus with a 2-amino-ethyl side
chain. Hill and Usherwood suggest that these analogues may
act in the insect preparation 'by blocking receptor sites for a
transmitter with a chemical structure just dissimilar enough to
prevent the 5-HT compounds substituting for it in excitation'.

5-HT has been found to reduce the amplitude of spontaneous
miniature potentials in various insect muscle preparations and
to block discharge completely at high concentration (Usherwood, 1963). These results suggest that 5-HT blocks transmission at the insect neuromuscular junction by postsynaptic
rather than presynaptic action.

At a lobster neuromuscular junction, however, application of 5-HT at relatively dilute concentrations appears to enhance transmitter action (Bergmann *et al.*, 1959). In this preparation the considerable increase in excitatory postsynaptic potentials caused by this amine was not accompanied by significant changes in the resting potential or membrane conductance.

5-HT has been demonstrated to have an excitatory effect on crustacean (Florey and Florey, 1954; Welsh, 1957) and insect heart preparations (Twarog, 1957; Davey, 1961; Colhoun, 1963a). Colhoun (1963a) also indicates that 5-HT is pharmacologically active when tested upon cockroach gut and malpighian tubules. With the heart of *Periplaneta* the threshold for 5-HT stimulation is about 10^{-8} M (Davey, 1961). In both crustacean and insect hearts the acceleratory action of 5-HT is greatly in excess of that exerted by acetylcholine, adrenalin or nor-adrenalin. Tissue extracts from *Carcinus* hearts have, however, indicated the presence of 6-HT, which has been shown to be ten times as active as 5-HT as a cardioaccelerator (Kerkut and Price, 1964).

Finally it should be mentioned that 5-HT appears to have an action on sensory nerve endings. Applications of relatively low concentrations of this substance to proprioceptors in the abdomen and legs of crayfish have been shown to cause these structures to discharge (Florey, 1954b).

It will be seen from the relatively small amount of information outlined above that 5-HT seems to exert a multiplicity of action in the peripheral nervous system of arthropods. This substance has been shown to exert an inhibitory effect at the neuromuscular junction in skeletal muscle, but to have an excitatory effect upon the heart and proprioceptors.

Gamma-aminobutyric acid and other amino acids

The role of gamma-aminobutyric acid (GABA) in the central nervous system of arthropods is of interest in view of the suggestion that this substance functions as an inhibitory chemical transmitter in the mammalian central nervous system. We have already seen that GABA occurs in arthropod ganglia (table 10), the concentration in the central nervous tissues of the insect species examined being relatively high.

Experiments on the inhibitory synapse of the crayfish nerve cord have shown that the application of GABA, at concentrations of above about 3.0×10^{-4} M, causes increased trans-membrane conductance, resulting in a pronounced depression of the excitatory postsynaptic potentials (Furshpan and Potter, 1959b). This amino acid produced no effect on the spontaneous activity of crayfish abdominal ganglion in the concentration range 10^{-10} to 10^{-2} M (Hichar, 1960). GABA has, however, been shown to have an inhibitory effect on the electrical activity of the nerve cord of the caterpillar *Dendrolinus pini* (Vereschtchagin, Sytinsky and Tyschenko, 1960), on the auditory synapses of the locust prothoracic ganglion (Suga and Katsuka, 1961) and on excitatory synapses in the terminal abdominal ganglion of the cockroach (Gahery and Boistel, 1965). The effects observed in these insect preparations were only obtained, however, following application of relatively high concentrations of GABA, in the range of 10^{-3} to 10^{-2} M. This limited amount of evidence would seem to indicate that GABA can affect both inhibitory and excitory synapses in the central nervous system.

Injection of the glutamic acid decarboxylase inhibitor, thiosemicarbazide, into the crayfish (Florey and Chapman, 1961) and the cockroach (Colhoun, unpublished) has been shown to be without effect, whereas in mammals this substance is known to produce convulsive effects (Killam, 1957). It is not known, however, to what extent the thiosemicarbazide affected the activity of the glutamic acid decarboxylase which has been demonstrated in arthropod ganglia (p. 86), and the level of GABA in the central nervous tissues in these experiments. This substance has, in fact, been found not to lower the GABA content in mouse brain tissues (Baltzer, Holtz and Palm, 1960), so that the lack of effect of thiosemicarbazide cannot be regarded as conclusive evidence that GABA is not an inhibitory substance in the crayfish central nervous system.

The effects of GABA on the peripheral nervous system have been intensively studied in some crustacean species, following the original investigation of Bazemore, Elliott and Florey (1957), and it has been found that this compound duplicates the effects of the neurally released inhibitory transmitter substance (cf. Kuffler, 1960; Florey, 1960). Much of this work has been concerned with the effects of GABA on the crustacean stretch receptor (see p. 11) described by Alexandrowicz (1951), which was first used as a physiological preparation by Wiersma, Furshpan and Florey (1953). Some of the effects of GABA on this preparation can be illustrated by the classical experiments of Kuffler and Edwards (1958). These authors showed that application of this substance to a stretched receptor caused a cessation of discharge and a repolarization of the membrane potential towards its resting level. The repolarization was shown to exhibit a graded response, which depended upon the concentration of GABA in the perfusion fluid. It has been shown that this substance exerts its effects upon the dendrites and cell body and does not affect axonal conduction. In the relaxed state the receptor cell becomes relatively insensitive to applied GABA. The inhibitory processes appear to be a result of an increased permeability of the postsynaptic membrane to potassium and/or chloride ions brought about by the transmitter molecules (cf. Kuffler, 1960).

Very little is known concerning the molecular basis for the action of GABA on the stretch receptor neurone. Comparison of the action of a series of related compounds has led to the suggestion that the inhibitory action results from the presence of both acidic and basic groups in the molecule (McGeer *et al.*, 1961). By analogy with the hypothesis erected for mammalian spinal neurones (Curtis and Watkins, 1960), it is suggested that the gap of 4 Å between the groups may correspond to the distance between receptor sites. The relative ineffectiveness of the branched or substituted derivatives of GABA (Edwards and Kuffler, 1959; McGeer *et al.*, 1961) would, according to this hypothesis, result from the interference of the side chains with the orientation with respect to the twin receptor sites. In crayfish muscle, however, recent observations suggest that different receptors are involved in producing the action of GABA and glutamate (Takeuchi and Takeuchi, 1965).

An attempt has been made to visualize the receptor sites involved in producing the inhibitory effect of GABA by incubating stretch receptors with ^{3}H-labelled solutions of the amino acid (Sisken and Roberts, 1964). The results demonstrated rapid binding of this compound and revealed a greater degree of localization at the axodendritic endings. The pattern of uptake of labelled GABA was very different from that of radioactive thymidine and leucine, and was found to be dependent upon the level of sodium ions in the bathing medium. The autoradiographic technique used in this investigation did not, however, give sufficient resolution to determine whether the binding of the GABA occurred at the pre- or postsynaptic membranes.

The effect of GABA on crayfish neuromuscular transmission has been shown to be localized at the inhibitory junction area (Takeuchi and Takeuchi, 1964). Application of the amino acid produces a postsynaptic conductance increase and also mimics the effects of stimulation of the inhibitory fibres by inhibiting the presynaptic excitatory axon, where it is presumed to reduce the release of transmitter substance (Dudel and Kuffler, 1961). An effect of GABA on the presynaptic membrane can also be deduced from the observation that the inhibitory effect of GABA is abolished by section of the inhibitory fibres or complete denervation, but is not affected by cutting the motor nerve when the inhibitory fibres are intact (Kuntsova, 1961). These results are in accord with the suggestion that the effects of GABA may be to alter the stability of the presynaptic membrane: decreasing the stability of 'synaptic' vesicles at inhibitory endings and increasing the stability at excitatory endings (Sisken and Roberts, 1964).

In view of the well-established effects of GABA on both central and peripheral synapses in the nervous system it is of obvious interest to know to what extent this substance can be identified with neurally released inhibitory compounds identified in arthropods. This question will be considered in the next section of this monograph.

Some other amino acids have also been demonstrated to effect synaptic transmission in arthropod species. Thus β-alanine has been shown to inhibit electrical activity of the nerve cord of the caterpillar of *Dendrolinus pini* (Vereshtchagin *et al.*, 1960) and to enhance spontaneous activity in crayfish ganglia at 10^{-4} M, producing inhibition at 10^{-2} M (Hichar, 1960). Glutamate has been found to be the only excitatory compound extracted from lobster nerve (Kravitz *et al.*, 1963). This amino acid, which occurs in high concentration in crustacean axons (table 10), imitates the excitatory transmitter in crayfish muscle (Takeuchi and Takeuchi, 1964) and has been shown to exert pharmacological effects upon the crayfish gut (Florey, 1961*a*; Jones, 1962) and stretch receptor (Elliott and Florey, 1956). It is at the moment impossible to identify glutamate as a neurally released transmitter, although it has been suggested that it may act as an excitatory compound in conjunction with the inhibitory GABA at the crustacean neuromuscular junction (Takeuchi and Takeuchi, 1965). The effects of the glutamate level in the blood (table 11) upon the excitability of synaptic membranes have not been determined.

Substance I

Both peripheral and central nervous tissues of decapod crustaceans have been found to contain a substance which exerts a pronounced inhibitory effect on crayfish stretch receptor neurones, heart ganglia and neuromuscular transmission (Florey, 1954 a, c, 1962). Florey (1960) has proposed that this substance be known as Substance I to distinguish it from mammalian Factor I, which appears to consist of a mixture of compounds and which also has an inhibitory action on crustacean stretch receptors.

Many of the effects of GABA on crustacean nervous structures, described in the preceding section, mimic those exerted by Substance I, and the question arises as to whether this inhibitory substance can be identified with the amino acid.

It has been found that Substance I is contained within the inhibitory fibres of crab peripheral nerve, but is absent from sensory and motor fibres (Florey and Biederman, 1960). The inhibitory fibres contain no detectable acetylcholine, although appreciable amounts occur in the sensory fibres (p. 94). The substance extracted from inhibitory fibres was identical in its action, on a variety of crustacean preparations, to both mammalian Factor I and to GABA. The inhibitory substance isolated from the cardio-inhibitory fibres of *Homarus americanus* also appears to be identical with Substance I (Florey, 1960).

The possibility that the inhibitory substance of crustacean nervous tissues is, in fact, GABA was strengthened by the observation that, in addition to duplicating the effects of the neurally released compound, its effects were abolished by picrotoxin, a substance which is known to block the action of inhibitory neurones in the crustacean stretch receptor (Elliott and Florey, 1956). Picrotoxin has also been found to block the action of GABA in crustacean neuromuscular transmission (Van der Kloot and Robbins, 1959; Reuben, Bergmann and Grundfest, 1959).

The concentration of GABA isolated from inhibitory nerves in *Cancer* and *Homarus* is, it is claimed, insufficient to account for the observed effect of Substance I (Florey, 1960; Florey and Chapman, 1961). More recent investigations have demonstrated, however, that GABA is the most active of the blocking agents extracted from peripheral and central nervous tissues of *Homarus* (Dudel *et al.*, 1963; Kravitz *et al.*, 1963). Kravitz *et al.* (1963) have, in addition, stated that GABA accounts for between 30 and 50% of the total blocking activity of the lobster nervous tissues, the remainder being largely contributed by taurine and betaine. These authors suggest that the latter substances might have contributed to the discrepancy between measured tissue concentration of GABA and the activity of the inhibitor substance claimed by Florey and his co-workers. They also conclude that GABA has a specific role which is linked with the function of inhibitory neurones. Florey and his co-workers still claim, however, that the inhibitory activity in extracts of crab peripheral nerve is about 120 times more than can be accounted for by the presence of GABA in the tissue (personal communication). It has also been claimed that the reversal potentials for the action of Substance I and GABA are closely similar, both resulting from changes in chloride permeability in crayfish stretch receptors (Iwasaki and Florey, unpublished).

It is at the moment impossible to identify conclusively the neurally released inhibitory compound. The possibility would seem to exist, however, that GABA may exert its effect by decreasing the stability of the presynaptic membrane (p. 117), the actual inhibitory substance being sequestered in the vesicular structures of the presynaptic cytoplasm (p. 11).

CHAPTER II

Concluding Remarks

I t is now relevant to consider in a general way to what extent the specialized structure of the nervous system and the peculiar composition of the body fluids of some species are reflected in equivalent specialization in the physiological organization of arthropod nervous tissues. In particular it is of interest to know how the chemical events associated with nervous activity are specialized in this phylum.

We have seen that the nervous tissues of the arthropod species examined are in a dynamic equilibrium with the ions and molecules of the body fluids. Most of the exchanges of water and inorganic ions taking place between the blood and the extracellular fluid of both peripheral and central nervous tissues appear to be largely passive in nature and only in the nervous system of a phytophagous insect has an active uptake of sodium ions been demonstrated. The secretion of this cation into the extracellular fluid thus maintains a high sodium environment around the nerve cells despite the very peculiar ionic composition of the blood of these insect species.

The mechanism for the transmission of nerve impulses in both crustacean and insect axons can be largely accounted for by the membrane theory, in which the resting potential is determined, as a first approximation, by the high intracellular potassium level relative to the surrounding medium, and the action potential by the influx of sodium ions from the high concentration in the extracellular fluid. The effects of changing the concentration of potassium in the bathing medium have shown, however, that the conductance of the membrane to other ions also contributes to the measured resting potential of arthropod axons. As with squid axons, the conductance of additional ions partly determines the extent of the action potential in insect axons, although in crayfish giant axons the magnitude of the action potential was similar to that which would be predicted for an ideal sodium electrode. Some quantitative differences which exist between the form of the action potential in arthropod species and those described from the classical experiments on squid giant axons can be related to specialization in the organization of arthropod nervous tissues. Thus, for example, the relatively rapid decline in the negative after-potential demonstrated in insect axons can be accounted for by the architecture of the adjacent extracellular system demonstrated with the electron-microscope.

The maintenance of a low intracellular sodium level and a high concentration of potassium in arthropod axons appears to be effected, as in excitable cells from other animal groups, by a linked sodium-potassium pump. From the limited amount of information available it appears that this linked ion pump shares some of the characteristics demonstrated in the squid axon membrane, sodium extrusion probably involving the utilization of high-energy phosphate compounds.

One significant difference between arthropod and vertebrate central nervous tissues is the relative accessibility of the former to water-soluble molecules. In insects this accessibility appears to result from the high permeability of the nerve sheath and from the very much larger extracellular volume demonstrated both physiologically and with the electron-microscope. It has been suggested that the vertebrate 'blood-brain barrier' phenomena may be merely a consequence of the extremely restricted extracellular system in vertebrate brain tissues (Edström, 1958). This interpretation implies that the underlying physiological processes involved in the exchange of ions and molecules between the blood and the tissues of the central nervous system may be fundamentally similar in arthropod and vertebrate animals, the apparent differences resulting from the variations in the organization of the extracellular system in these two groups of animals. The alternative explanation of the vertebrate 'blood-brain barrier' (i.e. that it results from a discrete diffusion barrier) would imply, on the other hand, the existence of fundamental differences in the exchange processes taking place between the blood and central nervous tissues in arthropod and vertebrate animals.

It is apparent from the preceding pages that much of the chemistry and physiology of synaptic transmission in arthropods remains obscure. The identification of transmitter substances at central and peripheral synapses has not, for example, been rigorously accomplished for any arthropod species. The available evidence would seem to indicate, however, that many of the basic cytological features of the synaptic regions in the central nervous tissues of this invertebrate group are essentially similar to those in the vertebrate nervous system. Synaptic transmission is thus visualized as being mediated by a chemical transmitter substance contained in vesicles in the presynaptic cytoplasm. In the case of the insect central nervous system there is strong circumstantial evidence that the transmitter contained in the vesicles is, in fact, acetylcholine. This substance does not, however, appear to be involved at neuromuscular synapses in arthropods. One notable feature of arthropod synapses is the relatively high concentrations of pharmacologically active substances which are frequently required to affect transmission processes. In the case of acetylcholine it has been suggested that this may be a reflexion of the relative insensitivity of the postsynaptic membrane to the transmitter substance. This degree of insensitivity has, in turn, been related to the extremely high concentrations of both acetylcholine and cholinesterase demonstrated in some arthropod nervous tissues. Similarly the high levels of γ-aminobutyric acid and catecholamines in these tissues also parallel the relatively high concentrations of these substances required to affect synaptic transmission processes. In insects, at least, it is conceivable that the degree of insensitivity of postsynaptic membranes might be a consequence of the extremely high amino acid concentration of the blood and extracellular fluids. These compounds are known to exert both inhibitory and excitatory effects on arthropod synapses and it might be expected, therefore, that the effective concentration of the transmitter substances would necessarily be correspondingly high in order to exceed the effects produced by the amino acids normally present in the body fluids.

It is to be hoped that future research may help to elucidate some of the many problems associated with the study of these mechanisms of synaptic transmission in arthropods.

REFERENCES

ABBOTT, B. C. (1958). The loss of potassium during stimulation of the limb nerve of *Maia*. Appendix to Abbott, Hill and Howarth (1958). *Proc. Roy. Soc.* B, **148**, 184–7.

ABBOTT, B. C., HILL, A. V. and HOWARTH, J. V. (1958). The positive and negative heat production associated with a nerve impulse. *Proc. Roy. Soc.* B, **148**, 149–87.

ABBOTT, B. C., HOWARTH, J. V. and RITCHIE, J. M. (1965). The initial heat production associated with the nerve impulse in crustacean and mammalian non-myelinated nerve fibres. *J. Physiol.* **178**, 368–83.

ADELMAN, W. J. jr. (1956). The effect of external calcium and magnesium depletion on single nerve fibers. *J. gen. Physiol.* **39**, 753–72.

ADELMAN, W. J. jr. and ADAMS, J. (1959). Effects of calcium lack on action potential of motor axons of the lobster limb. *J. gen. Physiol.* **42**, 655–64.

ADELMAN, W. J. jr. and DALTON, J. C. (1960). Interactions of calcium with sodium and potassium in membrane potentials of the lobster giant axon. *J. gen. Physiol.* **43**, 609–19.

ALEXANDROWICZ, J. S. (1951). Muscle receptor organs in the abdomen of *Homarus vulgaris* and *Palinurus vulgaris*. *Quart. J. micr. Sci.* **92**, 163–99.

ASHHURST, D. E. (1959). The connective-tissue sheath of the locust nervous system: a histochemical study. *Quart. J. micr. Sci.* **100**, 401–12.

ASHHURST, D. E. (1961 *a*). A histochemical study of the connective tissue sheath of the nervous system of *Periplaneta americana*. *Quart. J. micr. Sci.* **102**, 455–61.

ASHHURST, D. E. (1961 *b*). An acid mucopolysaccharide in cockroach ganglia. *Nature, Lond.*, **191**, 1224–5.

ASHHURST, D. E. (1961 *c*). The cytology and histochemistry of the neurones of *Periplaneta americana*. *Quart. J. micr. Sci.* **102**, 399–405.

ASHHURST, D. E. and CHAPMAN, J. A. (1961). The connective-tissue sheath of the nervous system of *Locusta migratoria*: an electron microscope study. *Quart. J. micr. Sci.* **102**, 463–7.

ASHHURST, D. E. and PATEL, N. G. (1963). Hyaluronic acid in cockroach ganglia. *Ann. ent. Soc. Amer.* **56**, 182–4.

ASHHURST, D. E. and RICHARDS, A. G. (1964). The histochemistry of the connective tissue associated with the central nervous system of the pupa of the wax moth, *Galleria mellonella* L. *J. Morph.* **114**, 237–46.

ASPEREN, K. VAN and ESCH, I. VAN (1956). The chemical composition of the haemolymph in *Periplaneta americana*, with special reference to the mineral constituents. *Arch. néerl. Zool.* **11**, 342–60.

AUGUSTINSSON, K-B. (1948). Cholinesterases. *Acta physiol. scand.* **15**, Suppl. 52, 1–182.

AUGUSTINSSON, K-B. and GRAHN, M. (1954). The occurrence of choline esters in the honey bee. *Arch. physiol. scand.* **32**, 174–90.

BACCETTI, B. (1956). Lo stroma di sostegno di organi degli insetti esaminato a luce polarizzata. *Redia*, **41**, 259–76.

BACCETTI, B. (1957). Observations by polarized light on the supporting stroma of some organs in insects. *Exp. Cell Res.* **13**, 158–60.

BACQ, Z. M. (1935). La choline-estérase chez les Invertébrés. L'insensibilité des Crustacés a l'acétylcholine. *Arch. int. Physiol.* **42**, 47–60.

BACQ, Z. M. (1947). L'acétylcholine et l'adrénaline chez les Invertébrés. *Biol. Rev.* **22**, 73–91.

BAKER, P. F. (1963). The relationship between phosphorus metabolism and the sodium pump in intact crab nerve. *Biochem. biophys. acta*, **75**, 287–9.

BAKER, P. F. (1965). Phosphorus metabolism of intact crab nerve and its relation to the active transport of ions. *J. Physiol.* **180**, 383–423.

BAKER, P. F., HODGKIN, A. L. and SHAW, T. I. (1962). The effects of changes in internal ionic concentrations on the electrical properties of perfused giant axons. *J. Physiol.* **164**, 355–74.

BALZER, H., HOLTZ, P. and PALM, D. (1960). Untersuchungen über die biochemischen Grundlagen der konvulsiven Wirkung von Hydraziden. *Arch. exp. Path. Pharm.* **239**, 520–52.

BARTON-BROWNE, L., DODSON, L. F., HODGSON, E. S. and KIRALY, J. K. (1961). Adrenergic properties of the cockroach corpus cardiacum. *Gen. comp. Endocrin.* **1**, 232–6.

BAYLOR, E. R. (1942). Cardiac pharmacology of the cladoceran, *Daphnia. Biol. Bull., Wood's Hole*, **83**, 165–72.

BAZEMORE, A. W., ELLIOTT, K. A. C. and FLOREY, E. (1957). Isolation of Factor I. *J. Neurochem.* **1**, 334–9.

BEAMS, H. W. and KING, R. L. (1932). Cytoplasmic structures in the ganglion cells of certain Orthoptera, with special reference to the Golgi bodies, mitochondria, 'vacuome', intracellular trabeculae (trophospongium), and neurofibrillae. *J. Morph.* **53**, 59–89.

BEAR, R. S. and SCHMITT, F. O. (1937). Optical properties of the axon sheaths of crustacean nerves. *J. cell. comp. Physiol.* **9**, 275–86.

BEKKER, J. M. and KRIJGSMAN, B. J. (1951). Physiological investigations into the heart function of *Daphnia. J. Physiol.* **115**, 249–57.

BERESINA, M. (1932). The resting heat production of nerve. *J. Physiol.* **76**, 170–80.

BERESINA, M. and FENG, T. P. (1933). The heat production of crustacean nerve. *J. Physiol.* **77**, 111–38.

BERGMANN, F., REUBEN, J. P. and GRUNDFEST, H. (1959). Action of biogenic amines and derivatives on lobster neuromuscular transmission. *Biol. Bull., Woods Hole*, **117**, 405.

BERMAN, R. (1954). Sulfhydryl groups of choline acetylase. *Biochim. biophys. acta*, **14**, 442–3.

BERTHOUMEYROUX, J. (1935). *Recherches sur le Pouvoir Réducteur des Liquides du Milieu Intérieur de Quelques Invertébrés Marins*. Imprimerie-Librairie de l'Université: Bordeaux.

BOCCACCI, M., NATALIZI, G. and BETTINI, S. (1960). Research on the mode of action of halogen containing thiol alkylating agents on insects. Effect of iodoacetic acid on choline acetylase. *J. ins. Physiol.* **4**, 20–6.

BOISTEL, J. (1960). *Caractéristiques Fonctionnelles des Fibres Nerveuses et des Récepteurs Tactiles et Olfactifs des Insectes.* Librairie Arnette : Paris.

BOYLE, P. J. and CONWAY, E. J. (1941). Potassium accumulation in muscle and associated changes. *J. Physiol.* **100**, 1–63.

BRANTE, G. (1949). Studies on lipids in the nervous system, with special reference to quantitative chemical determination and topical distribution. *Acta physiol. scand.* **18**, suppl. 63.

BRINK, F. (1954). The role of calcium ions in neural processes. *Pharm. Rev.* **6**, 243–98.

BRONK, D. W. (1939). Synaptic mechanisms in sympathetic ganglia. *J. Neurophysiol.* **2**, 380–401.

BULLOCK, T. H., GRUNDFEST, H., NACHMANSOHN, D. and ROTHENBER, M. A. (1947). Effect of di-isopropyl fluorphosphate (DFP) on action potential and cholinesterase of nerve, II. *J. Neurophysiol.* **10**, 63–78.

BULLOCK, T. H. and HORRIDGE, G. A. (1965). *Structure and Function in the Nervous System of Invertebrates.* Freeman : San Francisco and London.

BURKE, W., KATZ, B. and MACHNE, X. (1953). The effect of quaternary ammonium ions on crustacean nerve fibres. *J. Physiol.* **122**, 588–98.

CAJAL, S. R. Y and SANCHEZ, D. S. (1915). Contribución al conocimiento de los centros nerviosos de los insectos. Parte I. Retina y centros opticos. *Trab. Lab. Invest. biol. Univ. Madr.* **13**, 1–164.

CALDWELL, P. C., HODGKIN, A. L., KEYNES, R. D. and SHAW, T. I. (1960). The effects of injecting 'energy-rich' phosphate compounds on the active transport of ions in the giant axons of *Loligo*. *J. Physiol.* **152**, 561–90.

CAMERON, M. L. (1953). Secretion of an orthodiphenol in the corpus cardiacum of the insect. *Nature, Lond.*, **172**, 349–50.

CAMIEN, M. N., SARLET, H., DUCHÂTEAU, G. and FLORKIN, M. (1951). Non-protein amino acids in muscle and blood of some marine and freshwater Crustacea. *J. biol. Chem.* **193**, 881–5.

CANDY, D. J. and KILBY, B. A. (1959). Site and mode of trehalose bio-synthesis in the locust. *Nature, Lond.*, **183**, 1594–5.

CANDY, D. J. and KILBY, B. A. (1961). The biosynthesis of trehalose in locust fat body. *Biochem. J.* **78**, 531–6.

CARLISLE, D. B. (1956). An indole-alkylamine regulating heart beat in Crustacea. *Biochem. J.* **63**, 32.

CARLISLE, D. B. and KNOWLES, F. (1959). *Endocrine Control in Crustaceans.* Cambridge University Press.

CARRINGTON, C. B. and TENNEY, S. M. (1959). Chemical constituents of haemolymph and tissue in *Telea polyphemus* Gram. with particular reference to the question of ion binding. *J. ins. Physiol.* **3**, 402–13.

CARTA, S., FRONTALI, N. and VIVALDI, G. (1961). Il contenuto di amino-acidi liberi e altri composti nel tessuto nervose di *Apis mellifica* L. *R.C. Ist. sup. Sanit.* **24**, 407–22.

CHADWICK, L. E. (1963). Action on insects and other invertebrates. In *Handbuch der Experimentellen Pharmakologie Ergänzungswerk*, **15**, 741–98 (ed. G. B. Koelle). Springer : Berlin.

CHANG, S. C. and KEARNS, C. W. (1955). The occurrence of acetylcholine in the nerve tissue of American cockroaches. *Program 3rd Ann. Meeting entom. Soc. Amer.* 30–1.

125

CHEFURKA, W. and SMALLMAN, B. N. (1956). The occurrence of acetylcholine in the housefly, *Musca domestica* L. *Canad. J. Biochem.* **34**,
731–42.

CHENG, S. and WAELSCH, H. (1962). Carbon dioxide fixation in lobster
nerve. *Science*, **136**, 782–3.

COLE, W. H. (1940). The composition of fluids and sera of some marine
animals and of the sea in which they live. *J. gen. Physiol.* **23**, 575–84.

COLE, W. H. (1941). A perfusing solution for the lobster (*Homarus*) heart
and the effects of its constituent ions on the heart. *J. gen. Physiol.*
25, 1–6.

COLHOUN, E. H. (1958*a*). Physical release of acetylcholine from the thoracic
nerve cord of *Periplaneta americana*. *Nature, Lond.*, **181**, 490.

COLHOUN, E. H. (1958*b*). Acetylcholine in *Periplaneta americana*. I. Acetylcholine levels in nervous tissue. *J. ins. Physiol.* **2**, 108–16.

COLHOUN, E. H. (1958*c*). Acetylcholine in *Periplaneta americana* L. II.
Acetylcholine and nervous activity. *J. ins. Physiol.* **2**, 117–27.

COLHOUN, E. H. (1958*d*). Distribution of choline acetylase in insect con-
ductive tissue. *Nature, Lond.*, **182**, 1378.

COLHOUN, E. H. (1959). Physiological events in organophosphorus poisoning. *Can. J. Biochem. Physiol.* **37**, 1127–34.

COLHOUN, E. H. (1960). Acetylcholine in *Periplaneta americana* L. IV. Significance of esterase inhibition in intoxication, acetylcholine levels and
nervous conduction. *Can. J. Biochem. Physiol.* **38**, 1363–76.

COLHOUN, E. H. (1963*a*). Synthesis of 5-hydroxytryptamine in the American cockroach. *Experientia*, **19**, 9–10.

COLHOUN, E. H. (1963*b*). The physiological significance of acetylcholine in
insects and observations upon other pharmacologically active substances. In *Advances in Insect Physiology*, **1**, 1–46 (eds. J. W. L. Beament,
J. E. Treherne and V. B. Wigglesworth). Academic Press: London and
New York.

CORRIGAN, J. J. and KEARNS, C. W. (1963). Amino acid metabolism in
DDT poisoned American cockroaches. *J. ins. Physiol.* **9**, 1–12.

CORTEGGIANI, E. and SERFATY, A. (1939). Acétylcholine et cholinestérase
chez les insectes et les arachnides. *Compt. rendu. Soc. biol.* **131**, 1124–6.

COWAN, S. L. (1934). The action of potassium and other ions on the injury
potential and action current in *Maia* nerve. *Proc. Roy. Soc.* B, **115**,
216–60.

CROGHAN, P. C. (1958). The osmotic and ionic regulation of *Artemia salina*
(L.). *J. exp. Biol.* **35**, 219–33.

CROGHAN, P. C. (1959). The interstitial soil-water habitat and the evolution
of terrestrial arthropods. *Proc. R. phys. Soc. Edinb.* **27**, 103–4.

CURTIS, D. R. and WATKINS, J. C. (1960). Excitation of spinal neurones by
structurally related amino acids. *J. Neurochem.* **6**, 117–41.

CURTIS, H. J. and COLE, K. S. (1942). Membrane resting and action
potentials from the squid giant axon. *J. cell. comp. Physiol.* **19**, 135–44.

CZARNOWSKI, C. von (1954). Zur papierchromatographischen Blutzucker-
bestimmung bei der Honigbiene. *Naturwissenschaften*, **41**, 577.

DALTON, J. C. (1958). Effects of external ions on membrane potentials of
a lobster giant axon. *J. gen. Physiol.* **41**, 529–42.

DALTON, J. C. (1959). Effects of external ions on membrane potentials of a crayfish giant axon. *J. gen. Physiol.* **42**, 971–82.

DALTON, J. C. and ADELMAN, W. J. (1960). Some relations between action potential and resting potential of the lobster giant axon. *J. gen. Physiol.* **43**, 597–607.

DAUTERMAN, W. C., TALENS, A. and ASPEREN, K. VAN (1962). Partial purification and properties of flyhead cholinesterase. *J. ins. Physiol.* **8**, 1–14.

DAVEY, K. G. (1961). Substances controlling the rate of beating of the heart of *Periplaneta*. *Nature, Lond.*, **192**, 284.

DAVEY, K. G. (1964). The control of visceral muscles in insects. *Adv. in Insect Physiol.* **2**, 219–45.

DE ROBERTIS, E. (1958). Submicroscopic morphology and function of the synapse. *Exp. Cell Res.* Suppl. **5**, 347–69.

DE ROBERTIS, E. (1959). Submicroscopic morphology of the synapse. *Int. Rev. Cytol.* **8**, 61–96.

DE ROBERTIS, E. and SCHMITT, F. O. (1948). The effect of nerve degeneration on the structure of neurotubules. *J. cell. comp. Physiol.* **32**, 45–56.

DUCHÂTEAU, G. and FLORKIN, M. (1958). A survey of aminoacedemias with special reference to the high concentration of free amino acids in insect haemolymph. *Arch. int. Physiol. Biochem.* **66**, 573–91.

DUCHÂTEAU, G. and FLORKIN, M. (1959). Sur la trehalosémie des insectes et sa signification. *Arch. int. Physiol. Biochem.* **67**, 306–14.

DUCHÂTEAU, G., FLORKIN, M. and LECLERCQ, J. (1953). Concentrations des bases fixes et types de composition de la base totale de l'hémolymphe des insectes. *Arch. int. Physiol.* **61**, 518–49.

DUDEL, J., GRYDER, R., KAJI, A., KUFFLER, S. W. and POTTER, D. D. (1963). Gamma-aminobutyric acid and other blocking compounds in Crustacea. *J. Neurophysiol.* **26**, 721–8.

DUDEL, J. and KUFFLER, S. W. (1961). Presynaptic inhibition at the crayfish neuromusuclar junction. *J. Physiol.* **155**, 543–62.

EASTON, D. M. (1950). Synthesis of acetylcholine in crustacean nerve and nerve extract. *J. biol. Chem.* **185**, 813–16.

ECCLES, J. C. (1953). *The Neurophysiological Basis of Mind.* Clarendon Press, Oxford.

EDSTRÖM, R. (1958). An explanation of the blood-brain barrier phenomenon. *Acta psychiat. Kbh.* **33**, 403–16.

EDWARDS, C. and KUFFLER, S. W. (1959). Blocking effect of γ-aminobutyric acid and action of related compounds on a single nerve cell. *J. Neurochem.* **4**, 19–30.

EDWARDS, C., TERZUOLO, C. A. and WASHIZU, Y. (1963). The effect of changes in the ionic environment upon an isolated crustacean sensory neurone. *J. Neurophysiol.* **26**, 948–57.

EDWARDS, G. A. (1959). The fine structure of a multiterminal innervation of an insect muscle. *J. biophys. biochem. Cytol.* **5**, 241–4.

EDWARDS, G. A. (1960). Comparative studies on the fine structure of motor units. In *Vierter Internationaler Kongress für Elektronenmikroskopie*, **2**, 301–8. Springer: Berlin.

EDWARDS, G. A., RUSKA, H. and DE HARVEN, E. (1958a). Electron microscopy of peripheral nerves and neuromuscular junctions in the wasp leg. *J. biophys. biochem. Cytol.* **4**, 107–14.

EDWARDS, G. A., RUSKA, H. and DE HARVEN, E. (1958b). Neuromuscular junctions in flight and tymbal muscles of the cicada. *J. biophys. biochem. Cytol.* **4**, 251–6.

ELLIOTT, K. A. C. and FLOREY, E. (1956). Factor I—inhibiting factor from brain. Assay, condition in brain, stimulating and antagonizing substances. *J. Neurochem.* **1**, 181–92.

ELLIS, C. H., THIENES, C. H. and WIERSMA, C. A. G. (1942). The influence of certain drugs on the crustacean nerve-muscle system. *Biol. Bull., Wood's Hole*, **83**, 334–52.

EMMELIN, N. G. and MACINTOSH, F. C. (1952). Quoted by Eccles (1953).

ENGEL, G. L. and GERARD, R. W. (1935). The phosphorus metabolism of invertebrate nerve. *J. biol. Chem.* **112**, 379–92.

EULER, U. S. VON (1961). Occurrence of catecholamines in Acrania and invertebrates. *Nature, Lond.*, **190**, 170–1.

EYZAGUIRRE, C. and KUFFLER, S. W. (1955). Processes of excitation in the dendrites and in the soma of single isolated sensory nerve cells of the lobster and crayfish. *J. gen. Physiol.* **39**, 87–119.

FAIRBAIRN, D. (1958). Trehalose and glucose in helminths and other invertebrates. *Canad. J. Zool.* **36**, 787–95.

FATT, P. and GINSBORG, B. L. (1958). The ionic requirements for production of action potential in crustacean muscle fibres. *J. Physiol.* **142**, 516–43.

FATT, P. and KATZ, B. (1951). An analysis of the end-plate potential recorded with an intra-cellular electrode. *J. Physiol.* **115**, 320–70.

FATT, P. and KATZ, B. (1953). The electrical properties of crustacean muscle fibres. *J. Physiol.* **120**, 171–204.

FELDHERR, C. (1958). Physical properties of lobster nerve axoplasm. *Biol. Bull., Wood's Hole*, **115**, 328–9.

FENG, T. P. (1936). The heat production of nerve. *Ergebn. Physiol.* **38**, 71–132.

FENN, W. O. (1930). The anaerobic oxygen debt of frog nerve. *Amer. J. Physiol.* **92**, 349–61.

FENN, W. O., COBB, D. M., HEGNAUER, A. H. and MARSH, B. S. (1934). Electrolytes in nerve. *Amer. J. Physiol.* **110**, 74–96.

FLOREY, E. (1951). Neurohormone und Pharmakologie der Arthropoden. *Pflanzenschutzberichte*, Vienna, **7**, 81–141.

FLOREY, E. (1954a). Über die Wirkung von Acetylcholin, Adrenalin, Nor-Adrenalin, Faktor I und anderen Substanzen auf den isolierten Enddarm des Flusskrebses, *Cambarus clarki*. *Z. vergl. Physiol.* **36**, 1–8.

FLOREY, E. (1954b). Über die Wirkung von 5-Oxytryptamin (Enteramin) in der Krebsschere. *Z. Naturforsch.* **9**, 540–7.

FLOREY, E. (1954c). An inhibitory and an excitatory factor of mammalian central nervous system, and their action on a single sensory neuron. *Arch. int. Physiol.* **62**, 33–53.

FLOREY, E. (1960). Physiological evidence for naturally occurring inhibitory substances. In *Inhibition in the Nervous System and Gamma-Aminobutyric Acid* (eds. E. Roberts *et al.*). Pergamon: Oxford.

Florey, E. (1961a). A new test preparation for bio-assay of Factor I and gamma-aminobutyric acid. *J. Physiol.* **156**, 1–7.

Florey, E. (1961b). Comparative physiology: transmitter substances. *Ann. Rev. Physiol.* **23**, 501–28.

Florey, E. (1962). Comparative neurochemistry: inorganic ions, amino acids and possible transmitter substances of invertebrates. In *Neurochemistry*, 673–93 (eds. K. A. C. Elliott, I. H. Page and J. H. Quastel). Thomas: Springfield.

Florey, E. and Biederman, M. A. (1960). Studies on the distribution of Factor I and acetylcholine in crustacean peripheral nerve. *J. gen. Physiol.* **43**, 509–22.

Florey, E. and Chapman, D. D. (1961). The non-identity of the transmitter substance of crustacean inhibitory neurons and gamma-aminobutyric acid. *Comp. Biochem. Physiol.* **3**, 92–8.

Florey, E. and Florey, E. (1954). Über die mögliche Bedeutung von Enteramin (5-Oxy-Tryptamin) als nervoser Aktionssubstanz bei Cephalopoden und dekapoden Crustaceen. *Z. naturf.* **9**, 220.

Florkin, M. (1960). Blood Chemistry. In *The Physiology of Crustacea* (ed. T. H. Waterman). Academic Press: New York and London.

Foster, J. M. (1956). Enzymatic studies of mitochondria and other constituents of lobster and squid nerve fibres. *J. Neurochem.* **1**, 84–90.

Frankenhaeuser, B. and Hodgkin, A. L. (1956). The after-effects of impulses in the giant nerve fibres of *Loligo*. *J. Physiol.* **131**, 341–76.

Frankenhaeuser, B. and Hodgkin, A. L. (1957). The action of calcium on the electrical properties of squid axons. *J. Physiol.* **137**, 218–44.

Frontali, N. (1958). Acetylcholine synthesis in the housefly head. *J. ins. Physiol.* **1**, 319–26.

Frontali, N. (1961). Activity of glutamic acid decarboxylase in insect nerve tissue. *Nature, Lond.*, **191**, 178–9.

Frontali, N. (1964). Brain glutamic acid decarboxylase and synthesis of γ-aminobutyric acid in vertebrate and invertebrate species. In *Comparative Neurochemistry* (ed. D. Richter). Pergamon: Oxford.

Furshpan, E. J. and Potter, D. D. (1959a). Transmission at the giant motor synapses of the crayfish. *J. Physiol.* **145**, 289–325.

Furshpan, E. J. and Potter, D. D. (1959b). Slow post-synaptic potentials recorded from the giant motor fibre of the crayfish. *J. Physiol.* **145**, 326–35.

Gahery, Y. and Boistel, J. (1965). Study of some pharmacological substances which modify the electrical activity of the sixth abdominal ganglion of the cockroach (*Periplaneta americana*). In *The Physiology of the Insect Central Nervous System* (eds. J. E. Treherne and J. W. L. Beament). Academic Press: London and New York.

Geiger, A. (1962). Metabolism and function in the brain. In *Neurochemistry* (eds. K. A. C. Elliott, I. H. Page, and J. H. Quastel). Thomas: Springfield.

Gerard, R. W. (1927a). Studies on nerve metabolism. I. The influence of oxygen lack on heat production and action current. *J. Physiol.* **63**, 280–98.

GERARD, R. W. (1927b). Studies on nerve metabolism. II. Respiration in oxygen and nitrogen. *Amer. J. Physiol.* **82**, 381–404.

GEREN, B. B. and SCHMITT, F. O. (1954). The structure of the Schwann cell and its relation to the axon in certain invertebrate fibers. *Proc. nat. Acad. Sci., Wash.* **40**, 863–70.

GERSCH, M., FISCHER, F., UNGER, H. and KABITZA, W. (1961). Vorkommen von Serotonin im Nervensystem von *Periplaneta americana* L. *Z. Naturf.* **16**, 351–2.

GERSCH, M., UNGER, H. and FISCHER, F. (1957). Die Isolierung eines Neurohormones aus dem Nervensystem von *Periplaneta americana* und einige biologische Testverfahren. *Z. Friedrich-Schiller Universität, Jena,* **6**, 126–9.

GIACOBINI, E. (1965). Neurophysiological and biochemical correlations in isolated nerve cell preparations at rest and during activity. In *Drugs and enzymes (Proc. 2nd Int. Congr. Pharm.).* Pergamon Press: Oxford.

GILLES, R. (1962). Etude des relations existant entre la synthèse des acides aminés à partir d'oses et de dérivés d'oses et les phénomènes osmo-régulateurs au niveau de nerfs isolés de *Homarus vulgaris* L. *Ann. Soc. zool. Belg.* **92**, 191–3.

GILLES, R. and SCHOFFENIELS, E. (1964a). La synthèse des acides aminés de la chaine nerveuse ventrale du homard. *Biochim. biophys. acta,* **82**, 518–24.

GILLES, R. and SCHOFFENIELS, E. (1964b). Action de la vératrine de la cocaïne et de la stimulation électrique sur la synthèse et sur le pool des acides aminés de la chaine nerveuse ventrale du homard. *Biochim. biophys. acta,* **82**, 525–37.

GILMOUR, D. (1961). *Biochemistry of Insects.* Academic Press: New York and London.

GORDON, H. T. and WELSH, J. H. (1948). The role of ions in axon surface reaction to toxic inorganic compounds. *J. cell. comp. Physiol.* **31**, 395–415.

GRAMPP, W. and EDSTRÖM, J. E. (1963). The effect of nervous activity on ribonucleic acid in the crustacean receptor neuron. *J. Neurochem.* **10**, 725–31.

GRAVE, C. (1941). *Thesis Washington University, St Louis* (quoted in Richards, 1955).

GRAY, E. G. (1960). The fine structure of the insect ear. *Phil. Trans. B,* **243**, 77–94.

GREENBERG, M. J. (1960). Structure-activity relationships of tryptamine analogues on the heart of *Venus mercenaria. Brit. J. Pharmacol.* **15**, 375–88.

GUTTMAN, R. (1939). Stabilization of spider crab nerve membranes by alkaline earths, as manifested by resting potential measurements. *J. gen. Physiol.* **23**, 343–64.

HAMORI, J. (1961). Cholinesterase in insect muscle innervation with special reference to insecticide effects of DDT and DFP. *Bibl. Anat.* **2**, 194–206.

HARLOW, P. A. (1958). The action of drugs on the nervous system of the locust (*Locusta migratoria*). *Ann. appl. Biol.* **46**, 55–73.

HARRIS, E. J. (1954). Linkage of sodium and potassium active transport in human erythrocytes. *Symp. Soc. exp. Biol.* **8**, 228–41.

HEBB, C. O. and KRNJEVIĆ, K. (1962). The physiological significance of acetylcholine. In *Neurochemistry* (eds. K. A. C. Elliott, I. H. Page and J. H. Quastel). Thomas: Springfield.

HESLOP, J. P. and RAY, J. W. (1958). Phosphorus compounds of cockroach nerve and the effect of DDT. *Biochem. J.* **70**, 16.

HESLOP, J. P. and RAY, J. W. (1961). Nucleotides and other phosphorus compounds of the cockroach central nervous system. *J. ins. Physiol.* **7**, 127–40.

HESS, A. (1958). The fine structure of nerve cells and fibres, neurogli and sheaths of the ganglion chain in the cockroach (*Periplaneta americana*). *J. biophys. biochem. Cytol.* **4**, 731–42.

HEVESEY, G. and HAHN, L. (1941). Exchange of cellular potassium. *K. danske vidensk. Selsk., Biol. Medd.* **16**, 1–27.

HICHAR, J. K. (1960). Effects of γ-aminobutyric acid and picrotoxin on spontaneous activity in the central nervous system of the crayfish. *Nature, Lond.*, **188**, 1117–19.

HILL, A. V. (1928). The diffusion of oxygen and lactic acid through tissues. *Proc. Roy. Soc.* B, **104**, 39–96.

HILL, A. V. (1930). The heat production and recovery of crustacean nerve. *Proc. Roy. Soc.* B, **105**, 153–76.

HILL, A. V. (1932). *Chemical Wave Transmission in Nerve.* Cambridge University Press.

HILL, D. K. (1950). The effects of stimulation on the opacity of a crustacean nerve trunk and its relation to fibre diameter. *J. Physiol.* **111**, 283-303.

HILL, D. K. and KEYNES, R. D. (1949). Opacity changes in stimulated nerve. *J. Physiol.* **108**, 278–81.

HILL, P. R. and USHERWOOD, P. N. R. (1961). The action of 5-hydroxy-tryptamine and related compounds on neuromuscular transmission in the locust *Schistocerca gregaria*. *J. Physiol.* **157**, 393–401.

HINKE, J. A. M. (1961). The measurement of sodium and potassium activities in the squid axon by means of cation-selective glass micro-electrodes. *J. Physiol.* **156**, 314–35.

HODGKIN, A. L. (1947). The membrane resistance of a non-medullated nerve fibre. *J. Physiol.* **106**, 305–18.

HODGKIN, A. L. (1951). The ionic basis of electrical activity in nerve and muscle. *Biol. Rev.* **26**, 339–409.

HODGKIN, A. L. (1958). Ionic movements and electrical activity in giant nerve fibres. *Proc. Roy. Soc.* B, **148**, 1–37.

HODGKIN, A. L. (1964). *The Conduction of the Nervous Impulse.* Liverpool University Press.

HODGKIN, A. L. and HOROWICZ, P. (1959). Movements of Na and K in single muscle fibres. *J. Physiol.* **145**, 405–32.

HODGKIN, A. L. and HUXLEY, A. F. (1945). Resting and action potentials in single nerve fibres. *J. Physiol.* **104**, 176–95.

HODGKIN, A. L. and HUXLEY, A. F. (1947). Potassium leakage from an active nerve fibre. *J. Physiol.* **106**, 341–67.

HODGKIN, A. L. and KATZ, B. (1949). The effect of sodium ions on the electrical activity of the giant axon of the squid. *J. Physiol.* **108**, 37–77.

HODGKIN, A. L. and KEYNES, R. D. (1954). Movements of cations during recovery in nerve. *Symp. Soc. exp. Biol.* **8**, 423–37.

HODGKIN, A. L. and KEYNES, R. D. (1955). Active transport of cations in giant axons from *Sepia* and *Loligo*. *J. Physiol.* **128**, 28–60.

HODGKIN, A. L. and KEYNES, R. D. (1957). Movements of labelled calcium in squid giant axons. *J. Physiol.* **138**, 253–81.

HODGKIN, A. L. and RUSHTON, W. A. H. (1946). The electrical constants of a crustacean nerve fibre. *Proc. Roy. Soc.* B, **133**, 444–79.

HOLMES, E. G. (1929). Carbohydrates of crab nerve. *Biochem. J.* **23**, 1182–6.

HOLMES, E. G. and GERARD, R. W. (1929). Studies on nerve metabolism. IV. Carbohydrate metabolism of resting mammalian nerve. *Biochem. J.* **23**, 738–47.

HOLMES, W. (1942). The giant myelinated nerve fibres of the prawn. *Phil. Trans.* B, **231**, 293–312.

HOLMGREN, E. (1900). Weitere Mitteilung über die 'Saftkanälchen' der Nervenzellen. *Anat. Anz.* **18**, 149–53.

HORSTMANN, E. and MEVES, H. (1959). Die Feinstruktur des molekularen Rindengraves und ihre physiologisches Bedeutung. *Z. Zellforsch.* **49**, 569–604.

HOWDEN, G. F. and KILBY, B. A. (1956). Trehalose and trehalase in the locust. *Chem. and Ind.* 1453–4.

HOWDEN, G. F. and KILBY, B. A. (1961). Biochemical studies on insect haemolymph. II. The nature of the reducing material present. *J. ins. Physiol.* **6**, 85–95.

HOYLE, G. (1953). Potassium and insect nerve muscle. *J. exp. Biol.* **30**, 121–35.

HOYLE, G. (1954). Changes in the blood potassium concentration of the African migratory locust (*Locusta migratoria migratorioides*) during food deprivation and the effect on neuromuscular activity. *J. exp. Biol.* **31**, 260–70.

HOYLE, G. (1955). The anatomy and innervation of locust skeletal muscle. *Proc. Roy. Soc.* B, **143**, 281–92.

HOYLE, G. (1957). *Comparative Physiology of the Nervous Control of Muscular Contraction.* Cambridge University Press.

HUXLEY, T. H. (1880). *The Crayfish.* Kegan Paul: London.

IYATOMI, K. and KANESHINA, K. (1958). Localization of cholinesterase in the American cockroach. *Jap. J. appl. Ent. Zool.* **2**, 1–10.

JOHNSON, G. E. (1924). Giant nerve fibers in crustaceans with special reference to *Cambarus* and *Palaemonetes*. *J. comp. Neurol.* **36**, 323–73.

JONES, H. C. (1962). The action of L-glutamic acid and of structurally related compounds on the hind-gut of the crayfish. *J. Physiol.* **164**, 296–300.

JONES, J. C. (1964). The circulatory system of insects. In *Physiology of Insecta*, **3**, 1–107 (ed. M. Rockstein). Academic Press: New York and London.

JUDAH, J. D. and AHMED, K. (1964). The biochemistry of sodium transport. *Biol. Rev.* **39**, 160–93.

KANESHINA, K. (1961). Studies on the cholinesterases in insects, especially in relation to the mode of action of organophosphorus insecticides. *Bull.* 2, *Lab. appl. Ent., Faculty of Agr., Nagoya Univ., Japan.*

KATZ, B. (1936). Neuro-muscular transmission in crabs. *J. Physiol.* **87**, 199–200.

KATZ, B. (1962). The Croonian Lecture. The transmission of impulses from nerve to muscle, and the subcellular unit of synaptic action. *Proc. Roy. Soc.* B, **155**, 455–77.

KELLER, R. (1965). Die Aktivitäten von Enzymen der Glykolyse und des Citronensäurezyklus in den Organen von *Cambarus affinis* Say. *Z. vergl. Physiol.* **50**, 119–36.

KERKUT, G. A. and PRICE, M. A. (1964). Chromatographic separation of cardioaccelerators (6-HT and mucoprotein) from *Carcinus* heart. *Comp. Biochem. Physiol.* **11**, 45–52.

KEYL, M. J., MICHAELSON, I. A. and WHITTAKER, V. P. (1957). Physiologically active choline esters in certain marine gastropods and other invertebrates. *J. Physiol.* **139**, 434–54.

KEYNES, R. D. (1951). The leakage of radioactive potassium from stimulated nerve. *J. Physiol.* **113**, 99–114.

KEYNES, R. D. and LEWIS, P. R. (1951). The resting exchange of radioactive potassium in crab nerve. *J. Physiol.* **113**, 73–98.

KEYNES, R. D. and LEWIS, P. R. (1956). The intracellular calcium content of some invertebrate nerves. *J. Physiol.* **134**, 399–407.

KILBY, B. A. (1963). The biochemistry of the insect fat body. In *Advances in Insect Physiology*, **1**, 111–74 (eds. J. W. L. Beament, J. E. Treherne and V. B. Wigglesworth). Academic Press: London and New York.

KILLAM, K. F. (1957). Convulsant hydrazides. II: comparison of electrical changes and enzyme inhibition induced by the administration of thiosemicarbazide. *J. Pharmacol.* **119**, 263–71.

KINI, M. M. and QUASTEL, J. H. (1959). Carbohydrate-amino-acid interrelations in brain cortex *in vitro*. *Nature, Lond.*, **184**, 252–6.

KOEFOED-JOHNSEN, V. and USSING, H. H. (1953). The contribution of diffusion and flow to the passage of D_2O through living membranes. Effect of neurohypophyseal hormone on isolated anuran skin. *Acta physiol. scand.* **28**, 60–76.

KRAVITZ, E. A. (1962). Enzymatic formation of gamma-aminobutyric acid in the peripheral and central nervous system of lobsters. *J. Neurochem.* **9**, 363–70.

KRAVITZ, E. A., KUFFLER, S. W. and POTTER, D. D. (1963). Gamma-aminobutyric acid and other blocking compounds in Crustacea. III. Their relative concentration in separated motor and inhibitory axons. *J. Neurophysiol.* **26**, 739–51.

KRAVITZ, E. A., KUFFLER, S. W., POTTER, D. D. and GELDER, N. M. VAN (1963). Gamma-aminobutyric acid and other blocking compounds in Crustacea. II. Peripheral nervous system. *J. Neurophysiol.* **26**, 729–38.

KRAVITZ, E. A., POTTER, D. D. and GELDER, N. M. VAN (1962). Gamma-aminobutyric acid and other blocking substances extracted from crab muscle. *Nature, Lond.*, **194**, 382–3.

KRIJGSMAN, B. J. and KRIJGSMAN-BERGER, N. E. (1951). Physiological investigation into the heart function of arthropods. The heart of *Periplaneta americana*. *Bull. ent. Res.* **42**, 143–55.

133

Krnjević, K. (1955). The distribution of Na and K in cat nerves. *J. Physiol.* **128**, 473–88.

Kuffler, S. W. (1960). Excitation and inhibition in single nerve cells. In *The Harvey Lectures 1958–59*, 176–218. Academic Press: New York and London.

Kuffler, S. W. and Edwards, C. (1958). Mechanism of gamma-aminobutyric acid (GABA) action and its relation to synaptic inhibition. *J. Neurophysiol.* **21**, 589–610.

Kuntsova, M. Ya. (1961). The influence of γ-aminobutyric acid on the motor response of the normal and denervated adductor muscle of the crayfish. (In Russian.) *Bull. exp. Biol. Med., U.S.S.R.*, **52**, 8–12.

Lambremont, E. N. (1962). Enzymes in the boll weevil. I. Dehydrogenase of the brain and related structures. *J. ins. Physiol.* **8**, 181–90.

Levenbook, L. (1950). The composition of horse botfly (*Gasterophilus intestinalis*) larva blood. *Biochem. J.* **47**, 336–46.

Lewis, P. R. (1952). The free amino acids of invertebrate nerve. *Biochem. J.* **52**, 330–8.

Lewis, S. E. (1953). Acetylcholine in blowflies. *Nature, Lond.*, **172**, 1004–5.

Lewis, S. E. and Smallman, B. N. (1956). The estimation of acetylcholine in insects. *J. Physiol.* **134**, 241–56.

Lockwood, A. P. M. (1962). The osmoregulation of Crustacea. *Biol. Rev.* **37**, 257–305.

Lockwood, A. P. M. and Croghan, P. C. (1959). The composition of the haemolymph of *Petrobius maritimus*. *Nature, Lond.*, **184**, 370–1.

Long, C. (1961). *Biochemist's Handbook*. Spon: London.

Loomis, W. F. and Lipmann, F. (1948). Reversible inhibition of the coupling between phosphorylation and oxidation. *J. biol. Chem.* **173**, 807–8.

Lundberg, A. (1951). On the ability of some cations to inhibit the potassium depolarization of frog nerve fibres. *Acta physiol. scand.* **22**, 365–75.

Manery, J. (1939). Electrolytes in squid blood and muscle. *J. cell. comp. Physiol.* **14**, 365–9.

Margaria, R. (1931). The osmotic changes in some marine animals. *Proc. Roy. Soc. B*, **107**, 606–24.

Marnay, A. and Nachmansohn, D. (1937). Cholinestérase dans le muscle strié après dégénérescence du nerf moteur. *Compt. rendu, Soc. biol., Paris*, **126**, 785–9.

Maynard, D. M. (1961). Circulation and heart function. In *The Physiology of Crustacea* (ed. T. H. Waterman). Academic Press: New York and London.

Maynard, D. M. and Welsh, J. H. (1959). Neurohormones of the pericardial organs of brachyuran Crustacea. *J. Physiol.* **149**, 215–27.

Maynard, E. A. (1964). Esterases in crustacean nervous system. I. Electrophoretic studies in lobsters. *J. exp. Zool.* **157**, 251–66.

Maynard, E. A. and Maynard, D. M. (1960). Cholinesterases in the nervous system of the lobster, *Homarus americanus*. *Anat. Rec.* **137**, 350.

McColl, J. D. and Rossiter, R. J. (1950). Lipids of the nervous system of some invertebrates. *J. cell. comp. Physiol.* **36**, 241–50.

McGeer, E. G., McGeer, P. L. and McLennan, H. (1961). The inhibitory action of 3-hydroxytyramine, gamma-aminobutyric acid (GABA) and some other compounds towards the crayfish stretch receptor neuron. *J. Neurochem.* **8**, 36–49.

McIlwain, H. (1959). *Biochemistry and the Central Nervous System.* Churchill: London.

Mehrotra, K. N. (1961). Properties of choline acetylase from the housefly, *Musca domestica. J. ins. Physiol.* **6**, 215–21.

Meyerhof, O. and Schultz, W. (1929). Über die Atmung des marklosen Nerven. *Biochem. Z.* **206**, 158–70.

Michels, H. (1880). Beschreibung des Nervensystems von *Oryctes nasicornis* im Larven-, Puppen- und Käferzustande. *Z. wiss. Zool.* **34**, 611–700.

Mikalonis, S. J. and Brown, R. H. (1941). Acetylcholine and cholinesterase in the insect central nervous system. *J. cell. comp. Physiol.* **18**, 401–3.

Molloy, F. M. (1961). The histochemistry of the cholinesterase in the central nervous system of susceptible and resistant strains of the housefly, *Musca domestica* L., in relation to diazinon poisoning. *Bull. ent. Res.* **52**, 667–81.

Monnier, A. M. (1952). Properties of nerve axons (II). The damping factor as a functional criterion in nerve physiology. *Cold Spr. Harb. Symp. quant. Biol.* **17**, 69–95.

Monti, R. (1913). Sur les relations mutuelles entre les elements dans le système nerveux central des insectes. *Arch. Anat. Zool.* **15**, 349–433.

Moore, J. W. and Cole, K. S. (1960). Resting and action potentials of the squid giant axon *in vivo. J. gen. Physiol.* **43**, 961–70.

Mullins, L. J. (1956). The structure of nerve cell membranes. In *Molecular Structure and Functional Activity of Nerve Cells* (eds. R. G. Grenell and L. J. Mullins). American Institute of Biological Sciences: Washington.

Nachmansohn, D. (1938). Cholinestérase dans le tissu nerveux. *Compt. rendu Soc. biol., Paris,* **127**, 894–6.

Nachmansohn, D. and Rothenburg, M. A. (1945). Studies on cholinesterase. I. On the specificity of the enzyme in nerve tissue. *J. biol. Chem.* **158**, 653–66.

Naidu, M. B. (1955). Physiological action of drugs and insecticides on insects. *Bull. ent. Res.* **46**, 205–20.

Narahashi, T. (1961). Effect of barium ions on membrane potentials of cockroach giant axons. *J. Physiol.* **156**, 389–414.

Narahashi, T. (1963). The properties of insect axons. In *Advances in Insect Physiology,* **1**, 175–256 (eds. J. W. L. Beament, J. E. Treherne, and V. B. Wigglesworth). Academic Press: London and New York.

Narahashi, T. (1964). Restoration of action potential by anodal polarization in lobster giant axons. *J. cell. comp. Physiol.* **64**, 73–96.

Narahashi, T., Moore, J. W. and Scott, W. R. (1964). Tetrodotoxin blockage of sodium conductance increase in lobster giant axons. *J. gen. Physiol.* **47**, 965–74.

Narahashi, T., and Yamasaki, T. (1960). Mechanism of the after-potential production in the giant axons of the cockroach. *J. Physiol.* **151**, 75–88.

Nevis, A. H. (1958). Water transport in invertebrate peripheral nerve fibres. *J. gen. Physiol.* **41**, 927–58.

O'Brien, R. D. (1957). Esterases in the semi-intact cockroach. *Ann. ent. Soc. Amer.* **50**, 223–9.

O'Brien, R. D. and Fisher, R. W. (1958). The relation between ionization and toxicity to insects for some neuropharmacological compounds. *J. econ. Ent.* **51**, 169–75.

Östlund, E. (1954). The distribution of catecholamines in lower animals and their effect on the heart. *Acta physiol. scand.* **31**, suppl. 112, 1–67.

Overton, E. (1902). Beiträge zur allgemeinen Muskel- und Nervenphysiologie. *Pflüg. Arch. ges. Physiol.* **92**, 346–90.

Palay, S. L. (1958). The morphology of synapses in the central nervous system. *Exp. Cell Res.* suppl. **5**, 275–93.

Palay, S. L. and Palade, G. E. (1955). The fine structure of neurons. *J. biophys. biochem. Cytol.* **1**, 69–88.

Pappenheimer, J. R., Renkin, E. M. and Borrero, L. M. (1951). Filtration, diffusion and molecular sieving through peripheral capillary membranes. *Amer. J. Physiol.* **167**, 13–46.

Patterson, E. K., Dumm, M. E. and Richards, A. G. (1945). Lipids in the central nervous system of the honey bee. *Arch. Biochem.* **7**, 201–10.

Peterson, R. P. and Pepe, F. A. (1961). The fine structure of inhibitory synapses in the crayfish. *J. biophys. biochem. Cytol.* **11**, 157–69.

Pichon, Y. and Boistel, J. (1963). Modification of the ionic content of the haemolymph and of the activity of *Periplaneta americana* in relation to diet. *J. ins. Physiol.* **9**, 887–91.

Pipa, R. L. (1961). Studies on the hexapod nervous system. III. Histology and histochemistry of cockroach neuroglia. *J. comp. Neurol.* **116**, 15–22.

Pipa, R. L. and Cook, E. F. (1958). The structure and histochemistry of the connective tissue of sucking lice. *J. Morph.* **103**, 353–85.

Prosser, C. L. (1940a). Effects of salts upon 'spontaneous' activity in the nervous system of the crayfish. *J. cell. comp. Physiol.* **15**, 55–65.

Prosser, C. L. (1940b). Action potentials in the nervous system of crayfish. Effects of drugs and salts upon synaptic transmission. *J. cell. comp. Physiol.* **16**, 25–38.

Prosser, C. L. (1942). An analysis of the action of acetylcholine on hearts, particularly in arthropods. *Biol. Bull., Wood's Hole*, **83**, 145–64.

Prosser, C. L. (1943). An analysis of the action of salts upon abdominal ganglia of crayfish. *J. cell. comp. Physiol.* **22**, 131–45.

Prosser, C. L. and Brown, F. A. (1961). *Comparative Animal Physiology*, 2nd edn. Saunders: Philadelphia.

Ramsay, J. A. (1955). The excretory system of the stick insect *Dixippus morosus* (Orthoptera: Phasmidae). *J. exp. Biol.* **32**, 183–99.

Ray, J. W. (1964). The free amino acid pool of the cockroach (*Periplaneta americana*) central nervous system and the effect of insecticides. *J. ins. Physiol.* **10**, 587–97.

Ray, J. W. (1965). The free amino acid pool of cockroach (*Periplaneta americana*) central nervous system. In *The Physiology of the Insect Central Nervous System* (eds. J. E. Treherne and J. W. L. Beament). Academic Press: London and New York.

Reisberg, R. G. (1957). Properties and biological significance of choline acetylase. *Yale J. Biol. Med.* **29**, 403–35.

REUBEN, J. P., BERGMANN, F. and GRUNDFEST, H. (1959). Chemical excitation of presynaptic terminals at lobster neuromuscular junctions. *Biol. Bull.* **117**, 424.

REVEL, J. P., NAPOLITANO, L. and FAWCETT, D. W. (1960). Identification of glycogen in electronmicrographs of thin tissue sections. *J. biophys. biochem. Cytol.* **8**, 575–89.

RICHARDS, A. G. (1944). The structure of living insect nerves and nerve sheaths as deduced from the optical properties. *J. New York ent. Soc.* **52**, 285–310.

RICHARDS, A. G. and SCHNEIDER, D. (1958). Über den komplexen Bau der Membranen des Bindegewebes von Insekten. *Z. Naturf.* **13**, 680–7.

RIESNER, P. (1949). The protoplasmic viscosity of muscle and nerve. *Biol. Bull., Woods Hole,* **97**, 245–6.

ROBERTSON, J. D. (1939). The inorganic composition of the body fluids of three marine invertebrates. *J. exp. Biol.* **16**, 387–97.

ROBERTSON, J. D. (1953 a). Further studies on ionic regulation in marine invertebrates. *J. exp. Biol.* **30**, 277–96.

ROBERTSON, J. D. (1953 b). The ultrastructure of two vertebrate synapses. *Proc. Soc. exp. Biol. Med.* **82**, 219–23.

ROBERTSON, J. D. (1960). Osmotic and ionic regulation. In *The Physiology of Crustacea* (ed. T. H. Waterman). Academic Press: New York.

ROBERTSON, J. D. (1961). Ultrastructure of excitable membranes and the crayfish median-giant synapse. *Ann. New York Acad. Sci.* **94**, 339–89.

ROBERTSON, J. D. (1964). Unit membranes: a review with recent new studies of experimental alterations and a new subunit structure in synaptic membranes. In *Cellular Membranes in Development* (ed. M. Locke). Academic Press: New York and London.

ROEDER, K. D. (1948). The effect of anticholinesterases and related substances on nervous activity in the cockroach. *Johns Hopkins Hosp. Bull.* **83**, 587–99.

ROEDER, K. D. (1953). Electrical activity in nerves and ganglia. In *Insect Physiology* (ed. K. D. Roeder). Wiley: New York.

ROEDER, K. D. and WEIANT, E. A. (1950). The electrical and mechanical events of neuro-muscular transmission in the cockroach, *Periplaneta americana* L. *J. exp. Biol.* **27**, 1–13.

ROSS, L. S. (1915). The trophospongium of the nerve cell of the crayfish (*Cambarus*). *J. comp. Neurol.* **25**, 523–30.

ROTHENBERG, M. A. (1950). Studies on permeability in relation to nerve function. II. Ionic movements across axonal membranes. *Biochem. biophys. acta,* **4**, 96–114.

RUDALL, K. M. (1955). The distribution of collagen and chitin. *Symp. Soc. exp. Biol.* **9**, 49–71.

RUDENBERG, F. M. (1954). Preliminary experiments on the state of calcium in lobster nerve. *Fed. Proc.* **13**, 122.

RUSHTON, W. A. H. (1962). Nerve fibres. In *Principles of Human Physiology,* 13th edn (eds. H. Davson and G. Eggleton). Churchill: London.

SAKTOR, B. (1955). Cell structure and the metabolism of insect flight muscle. *J. biophys. biochem. Cytol.* **1**, 29–46.

SALACH, J. (1957). Quoted in Tobias (1958).

SCHALLEK, W. (1945). Action of potassium on bound acetylcholine in lobster nerve cord. *J. cell. comp. Physiol.* **26**, 15–24.

SCHALLEK, W. (1949). The glycogen content of some invertebrate nerves. *Biol. Bull., Wood's Hole*, **97**, 252–3.

SCHALLEK, W. and WIERSMA, C. A. G. (1948). The influence of various drugs on a crustacean synapse. *J. cell. comp. Physiol.* **31**, 35–47.

SCHALLEK, W. and WIERSMA, C. A. G. (1949). Effect of anti-cholinesterases on synaptic transmission in the crayfish. *Physiol. Comparata et Oecol.* **1**, 63–7.

SCHALLEK, W., WIERSMA, C. A. G. and ALLES, G. A. (1948). Blocking and protecting action of amines and ammonium compounds on a crustacean synapse. *Proc. Soc. exp. Biol. Med.* **68**, 174–8.

SCHARRER, B. C. J. (1939). The differentiation between neuroglia and connective tissue sheath in the cockroach (*Periplaneta americana*). *J. comp. Neurol.* **70**, 77–88.

SCHARRER, E. (1964). A specialized trophospongium in large neurons of *Leptodora* (Crustacea). *Z. Zellforsch.* **61**, 803–12.

SCHLIEPER, C. (1929). Über die Einwirkung niederer Salzkonzentrationen auf marine Organismen. *Z. vergl. Physiol.* **9**, 478–514.

SHANES, A. M. (1946). A neglected factor in studies of potassium distribution in relation to resting potential of nerve. *J. cell. comp. Physiol.* **27**, 115–18.

SHANES, A. M. (1949). Electrical phenomena in nerve. II. Crab nerve. *J. gen. Physiol.* **33**, 75–102.

SHANES, A. M. (1950). Potassium retention in crab nerve. *J. gen. Physiol.* **33**, 643–9.

SHANES, A. M. (1958a). Electrical aspects of physiological and pharmacological action in excitable cells. Part I. The resting cell and its alteration by extrinsic factors. *Pharm. Rev.* **10**, 59–164.

SHANES, A. M. (1958b). Electrochemical aspects of physiological and pharmacological action in excitable cells. Part II. The action potential and excitation. *Pharm. Rev.* **10**, 165–273.

SHANES, A. M. (1964). Movement of calcium in muscle and nerve. In *The Transfer of Calcium and Strontium Across Biological Membranes* (ed. R. H. Wasserman). Academic Press: New York and London.

SHANES, A. M., FREYGANG, W. H., GRUNDFEST, H. and AMATNIEK, E. (1959). Anaesthetic and calcium action in the voltage clamped squid giant axon. *J. gen. Physiol.* **42**, 793–802.

SHANES, A. M. and HOPKINS, H. S. (1948). The effect of potassium on 'resting potential' and respiration of crab nerve. *J. Neurophysiol.* **11**, 331–42.

SHAW, J. and STOBBART, R. H. (1963). Osmotic and ionic regulation in insects. In *Advances in Insect Physiology*, **1**, 315–99 (eds. J. W. L. Beament, J. E. Treherne and V. B. Wigglesworth). Academic Press: London and New York.

SHIGEMATSU, H. (1956). On the glycolytic and oxidative action by the silkworm, *Bombyx mori*. *J. seric. Sci., Tokyo*, **28**, 115–21.

SISKEN, B. and ROBERTS, E. (1964). Radioautographic studies of binding of γ-aminobutyric acid to the abdominal stretch receptor of the crayfish. *Biochem. Pharmacol.* **13**, 95–103.

Skou, J. C. (1957). The influence of some cations on an adenosinetri-phosphatase from peripheral nerves. *Biochem. biophys. acta*, **23**, 394–401.

Smallman, B. N. (1956). Mechanism of acetylcholine synthesis in the blowfly. *J. Physiol.* **132**, 343–57.

Smallman, B. N. (1958). The choline acetylase activity of rabbit brain. *J. Neurochem.* **2**, 119–27.

Smallman, B. N. (1961). Determination of choline acetylase activity. *Meth. med. Res.* **9**, 203–7.

Smallman, B. N. and Pal, R. (1957). The activity and intracellular distribution of choline acetylase in insect nervous tissue. *Bull. ent. Soc. Amer.* **3**, 25.

Smith, C. C. and Glick, D. (1939). Some observations on cholinesterase in invertebrates. *Biol. Bull. Woods Hole*, **77**, 321–2.

Smith, D. S. (1960). The innervation of the fibrillar flight muscle of an insect: *Tenebrio molitor*. *J. biophys. biochem. Cytol.* **8**, 447–66.

Smith, D. S. (1961). The organization of the flight muscle in a dragonfly *Aeschna* sp. *J. biophys. biochem. Cytol.* **11**, 119–44.

Smith, D. S. and Treherne, J. E. (1963). Functional aspects of the organization of the insect nervous system. In *Advances in Insect Physiology*, **1**, 401–84 (eds. J. W. L. Beament, J. E. Treherne and V. B. Wigglesworth). Academic Press: London and New York.

Smith, D. S. and Treherne, J. E. (1965). Electron microscope localization of acetylcholinesterase activity in the central nervous system of an insect (*Periplaneta americana*). *J. Cell. Biol.* **26**, 445–65.

Smith, D. S. and Wigglesworth, V. B. (1959). The occurrence of collagen in the perilemma of insect nerve. *Nature, Lond.*, **183**, 127–8.

Smith, R. I. (1939). Acetylcholine in the nervous tissues and blood of crayfish. *J. cell. comp. Physiol.* **13**, 335–44.

Spencer, W. S. (1956). *Handbook of Biological Data.* Saunders: Philadelphia and London.

Steele, J. E. (1963). The site of action of insect hyperglycaemic hormone. *Gen. Comp. Endocrinol.* **3**, 46–52.

Stegwee, D. (1960). The role of esterase in tetraethylpyrophosphate poisoning in the house-fly, *Musca domestica* L. *Canad. J. Biochem. Physiol.* **38**, 1417–30.

Steinbach, H. B. and Spiegelman, S. (1943). The sodium and potassium balance in squid axoplasm. *J. cell. comp. Physiol.* **22**, 187–96.

Stobbart, R. H. (1959). Studies on the exchange and regulation of sodium in the larva of *Aedes aegypti*. I. The steady-state exchange. *J. exp. Biol.* **36**, 641–53.

Straub, W. (1900). Zur Muskelphysiologie des Regenwurms. *Pflügers. Arch. ges. Physiol.* **79**, 379–99.

Suga, N. and Katsuki, Y. (1961). Pharmacological studies on the auditory synapse in a grasshopper. *J. exp. Biol.* **38**, 759–70.

Sutcliffe, D. W. (1963). The chemical composition of the haemolymph in insects and some other arthropods, in relation to their phylogeny. *Comp. Biochem. Physiol.* **9**, 121–35.

Takeuchi, A. and Takeuchi, N. (1964). The effect on crayfish muscle of iontophoretically applied glutamate. *J. Physiol.* **170**, 296–317.

Takeuchi, A. and Takeuchi, N. (1965). Localized action of gamma-aminobutyric acid on the crayfish muscle. *J. Physiol.* **177**, 225–38.

Tallan, H. H. (1962). A survey of the amino acids and related compounds in nervous tissue. In *Amino Acid Pools* (ed. J. T. Holden). Elsevier: Amsterdam.

Tipton, S. R. (1934). The calcium content of frog nerve. *Amer. J. Physiol.* **109**, 457–66.

Tobias, J. M. (1948*a*). Potassium, sodium and water interchange in irritable tissues and haemolymph of an omnivorous insect, *Periplaneta americana*. *J. cell. comp. Physiol.* **31**, 125–42.

Tobias, J. M. (1948*b*). The high potassium and low sodium in the body fluid of a phytophagous insect, the silkworm *Bombyx mori*, and the change before pupation. *J. cell. comp. Physiol.* **31**, 143–8.

Tobias, J. M. (1958). Experimentally altered structure related to function in the lobster axon with an extrapolation to molecular mechanisms in excitation. *J. cell. comp. Physiol.* **52**, 89–107.

Tobias, J. M. (1960). Further studies on the nature of the excitable system in nerve. I. Voltage induced axoplasm movements in squid axons. II. Penetration of surviving, excitable axons by proteases. III. Effects of proteases and of phospholipases on lobster giant axons resistance and capacity. *J. gen. Physiol.* **43**, 57–71.

Tobias, J. M., Agin, D. P. and Pawlowski, R. (1962). Phospholipid-cholesterol membrane model. Control of resistance by ions of current flow. *J. gen. Physiol.* **45**, 989–1001.

Tobias, J. M. and Bryant, S. H. (1955). An isolated giant axon preparation from the lobster nerve cord. *J. cell. comp. Physiol.* **46**, 163–82.

Tobias, J. M., Kollros, J. J. and Savit, J. (1946). Acetylcholine and related substances in the cockroach, fly and crayfish and the effect of DDT. *J. cell. comp. Physiol.* **28**, 159–82.

Travis, D. L. (1955). The moulting cycle of the spiny lobster *Panulirus argus* Latreille. III. Physiological changes which occur in the blood and urine during the moulting cycle. *Biol. Bull., Wood's Hole*, **109**, 484–503.

Treherne, J. E. (1954). The exchange of labelled sodium in the larva of *Aedes aegypti* L. *J. exp. Biol.* **31**, 386–401.

Treherne, J. E. (1958*a*). The absorption of glucose from the alimentary canal of the locust, *Schistocerca gregaria* (Forsk.). *J. exp. Biol.* **35**, 297–306.

Treherne, J. E. (1958*b*). The absorption and metabolism of some sugars in the locust, *Schistocerca gregaria* (Forsk.). *J. exp. Biol.* **35**, 611–25.

Treherne, J. E. (1960). The nutrition of the central nervous system in the cockroach, *Periplaneta americana*. The exchange and metabolism of sugars. *J. exp. Biol.* **37**, 513–33.

Treherne, J. E. (1961*a*). Sodium and potassium fluxes in the abdominal nerve cord of the cockroach, *Periplaneta americana*. *J. exp. Biol.* **38**, 315–22.

Treherne, J. E. (1961*b*). The movements of sodium ions in the isolated abdominal nerve cord of the cockroach, *Periplaneta americana*. *J. exp. Biol.* **38**, 629–36.

TREHERNE, J. E. (1961c). Exchanges of sodium ions in the central nervous system of an insect. *Nature, Lond.*, **191**, 1223–4.

TREHERNE, J. E. (1961d). The efflux of sodium ions from the last abdominal ganglion of the cockroach, *Periplaneta americana*. *J. exp. Biol.* **38**, 729–36.

TREHERNE, J. E. (1961e). The kinetics of sodium transfer in the central nervous system of the cockroach, *Periplaneta americana*. *J. exp. Biol.* **38**, 737–46.

TREHERNE, J. E. (1962a). Distribution of water and inorganic ions in the central nervous system of an insect (*Periplaneta americana*). *Nature, Lond.*, **193**, 750–2.

TREHERNE, J. E. (1962b). The distribution and exchange of some ions and molecules in the central nervous system of *Periplaneta americana*. *J. exp. Biol.* **39**, 193–217.

TREHERNE, J. E. (1962c). Some effects of the ionic composition of the extracellular fluid on the electrical activity of the cockroach abdominal nerve cord. *J. exp. Biol.* **39**, 631–41.

TREHERNE, J. E. (1962d). Transfer of substances between the blood and central nervous system in vertebrate and invertebrate animals. *Nature, Lond.*, **196**, 1181–3.

TREHERNE, J. E. (1965a). Some preliminary observations on the effects of cations on conduction processes in the abdominal nerve cord of the stick insect, *Carausius morosus*. *J. exp. Biol.* **42**, 1–6.

TREHERNE, J. E. (1965b). The distribution and exchange of inorganic ions in the central nervous system of the stick insect, *Carausius morosus*. *J. exp. Biol.* **42**, 7–27.

TREHERNE, J. E. (1965c). The comparative physiology of the transfer of substances between the blood and central nervous system. In *Studies in Comparative Biochemistry*, 81–106 (ed. K. A. Munday). Pergamon: Oxford.

TREHERNE, J. E. (1965d). The chemical environment of the insect central nervous system. In *The Physiology of the Insect Central Nervous System* (eds. J. E. Treherne and J. W. L. Beament). Academic Press: London.

TREHERNE, J. E. and SMITH, D. S. (1965a). The penetration of acetylcholine into the central nervous tissues of an insect (*Periplaneta americana*). *J. exp. Biol.* **43**, 13–21.

TREHERNE, J. E. and SMITH, D. S. (1965b). The metabolism of acetylcholine in the intact central nervous system of an insect (*Periplaneta americana* L.). *J. exp. Biol.* **43**, 441–54.

TRUJILLO-CENÓZ, O. (1959). Study on the fine structure of the central nervous system of *Pholus labruscoe* (Lepidoptera). *Z. Zellforsch.* **49**, 432–46.

TRUJILLO-CENÓZ, O. (1962). Some aspects of the structural organization of the arthropod ganglion. *Z. Zellforsch.* **56**, 649–82.

TURNER, R. S., HAGINS, W. A. and MOORE, A. R. (1950). Influence of certain neurotropic substances on central and synaptic transmission in *Callianassa*. *Proc. Soc. exp. Biol. Med.* **73**, 156–8.

TWAROG, B. M. (1957). Quoted in Twarog and Roeder (1957).

Twarog, B. M. and Roeder, K. D. (1956). Properties of the connective tissue sheath of the cockroach abdominal nerve cord. *Biol. Bull.*, *Woods Hole*, **111**, 278–86.

Twarog, B. M. and Roeder, K. D. (1957). Pharmacological observations on the desheathed last abdominal ganglion of the cockroach. *Ann. ent. Soc. Amer.* **50**, 231–7.

Ungar, G. and Romano, D. V. (1962). Fluorescence changes in nerve induced by stimulation. Their relation to protein configuration. *J. gen. Physiol.* **46**, 267–75.

Usherwood, P. N. R. (1963). Spontaneous miniature potentials from insect muscle fibres. *J. Physiol.* **169**, 149–60.

Van der Kloot, W. G. (1955). The control of neurosecretion and diapause by physiological changes in the brain of the cecropia silkworm. *Biol. Bull.*, *Wood's Hole*, **109**, 276–94.

Van der Kloot, W. G. and Robbins, J. (1959). The effects of γ-aminobutyric acid and picrotoxin on the junctional potential and the contraction of crayfish muscle. *Experientia*, **15**, 35–6.

Van Harreveld, A. (1939). The nerve supply of doubly and triply innervated crayfish muscles related to their function. *J. comp. Neurol.* **70**, 267–84.

Van Harreveld, A. and Crowell, J. (1964). Extracellular space in central nervous tissue. *Fed. Proc.* **23**, no. 2.

Vereshtchagin, S. M., Sytinsky, I. A. and Tyschenko, V. P. (1960). The effect of γ-aminobutyric acid and B alanine on bioelectrical activity of nerve ganglia of the pine moth caterpillar (*Dendrolimus pini*). *J. ins. Physiol.* **6**, 21–5.

Villegas, R. and Villegas, G. M. (1960). Characterization of the membranes in the giant nerve fibre of the squid. *J. gen. Physiol.* **43**, no. 5, suppl. 73.

Vonk, H. J. (1960). Digestion and metabolism. In *The Physiology of Crustacea* (ed. T. H. Waterman). Academic Press: New York and London.

Walop, J. N. (1950). Acetylcholine formation in the central nervous system of *Carcinus maenas*. *Acta physiol. pharm. néerl.* **1**, 333–5.

Walop, J. N. (1951). Studies on acetylcholine in the crustacean central nervous system. *Arch. int. Physiol.* **59**, 145–56.

Walop, J. N. and Boot, L. M. (1950). Studies on cholinesterase in *Carcinus maenas*. *Biochem. biophys. acta*, **4**, 566–71.

Wang, J. H., Robinson, C. V. and Edelman, I. S. (1953). Self-diffusion and structure of liquid water. III. Measurements of the self-diffusion of liquid water with H^2, H^3 and O^{18} tracers. *J. Amer. chem. Soc.* **75**, 466–70.

Wantanabe, A. and Grundfest, H. (1961). Impulse propagation at the septal and commissural junctions of crayfish lateral giant axons. *J. gen. Physiol.* **45**, 267–308.

Webb, D. A. (1940). Ionic regulation in *Carcinus maenas*. *Proc. Roy. Soc. B*, **129**, 107–36.

Welsh, J. H. (1957). Serotonin as a possible neurohumoral agent: evidence obtained in lower animals. *Ann. N.Y. Acad. Sci.* **66**, 618–30.

WELSH, J. H. (1961). Neurohumors and neurosecretion. In *The Physiology of Crustacea* (ed. T. H. Waterman). Academic Press: New York and London.

WELSH, J. H. and MOORHEAD, M. (1960). The quantitative distribution of 5-hydroxytryptamine in the invertebrates, especially in their nervous systems. *J. Neurochem.* **6**, 146–9.

WERMAN, R. and GRUNDFEST, H. (1961). Graded or all-or-none electrogenesis in arthropod muscle. II. The effect of alkali-earth and onium ions on lobster muscle fibres. *J. gen. Physiol.* **44**, 997–1027.

WERMAN, R., McCANN, F. V. and GRUNDFEST, H. (1961). Graded or all-or-none electrogenesis in arthropod muscle. I. The effects of alkali-earth cations on the neuromuscular system of *Romalia microptera*. *J. gen. Physiol.* **44**, 979–95.

WHITTAKER, V. P. (1959). The isolation and characterization of acetylcholine-containing particles from brain. *Biochem. J.* **72**, 694–706.

WIERSMA, C. A. G. (1941). The inhibitory nerve supply of the leg muscles of different decapod crustaceans. *J. comp. Neurol.* **74**, 63–79.

WIERSMA, C. A. G. (1947). Giant fiber system of the crayfish. A contribution to comparative physiology of synapses. *J. Neurophysiol.* **10**, 23–38.

WIERSMA, C. A. G. (1961). The neuromuscular system. In *The Physiology of Crustacea* (ed. T. H. Waterman). Academic Press: New York and London.

WIERSMA, C. A. G., FURSHPAN, E. and FLOREY, E. (1953). Physiological and pharmacological observations on muscle receptor organs of the crayfish, *Cambarus clarkii* Girard. *J. exp. Biol.* **30**, 136–50.

WIGGLESWORTH, V. B. (1942). The storage of protein, fat, glycogen and uric acid in the fat body and other tissues of mosquito larvae. *J. exp. Biol.* **19**, 56–77.

WIGGLESWORTH, V. B. (1956). The haemocytes and connective tissue formation in an insect, *Rhodnius prolixus* (Hemiptera). *Quart. J. micr. Sci.* **97**, 89–98.

WIGGLESWORTH, V. B. (1958). The distribution of esterase in the nervous system and other tissues of the insect, *Rhodnius prolixus*. *Quart. J. micr. Sci.* **99**, 441–50.

WIGGLESWORTH, V. B. (1959). The histology of the nervous system of an insect *Rhodnius prolixus* (Hemiptera). II. The central ganglia. *Quart. J. micr. Sci.* **100**, 299–313.

WIGGLESWORTH, V. B. (1960). The nutrition of the central nervous system of the cockroach, *Periplaneta americana*. The mobilization of reserves. *J. exp. Biol.* **37**, 500–12.

WILBRANDT, W. (1937). The effect of organic ions on the membrane potential of nerves. *J. gen. Physiol.* **20**, 519–41.

WINTON, M. Y., METCALF, R. L. and FUKUTO, T. R. (1958). The use of acetyl thiocholine in the histochemical study of the action of organophosphorus insecticides. *Ann. ent. Soc. Amer.* **51**, 436–41.

WOOD, D. W. (1957). The effects of ions upon neuromuscular transmission in a herbivorous insect. *J. Physiol.* **138**, 119–39.

WRIGHT, E. B. and TOMITA, T. (1965). A study of the crustacean axon repetitive response. II. The effects of cations, sodium, calcium (magnesium), potassium and hydrogen (pH) in the external medium. *J. cell. comp. Physiol.* **65**, 211–28.

WYATT, G. R. (1961). The biochemistry of insect haemolymph. *Ann. Rev. Ent.* **6**, 75–102.

WYATT, G. R. and KALF, G. F. (1956). Trehalose in insects. *Fed. Proc.* **15**, 388.

WYATT, G. R. and KALF, G. F. (1957). The chemistry of insect haemolymph. II. Trehalose and other carbohydrates. *J. gen. Physiol.* **40**, 833–47.

YAMASAKI, T. and NARAHASHI, T. (1957). Effects of oxygen lack, metabolic inhibitors and DDT on the resting potential of insect nerve. Studies on the mechanism of action of insecticides, XII. *Botyu-Kagaku*, **22**, 250–76.

YAMASAKI, T. and NARAHASHI, T. (1959a). The effects of potassium and sodium ions on the resting and action potentials of the cockroach giant axon. *J. ins. Physiol.* **3**, 146–58.

YAMASAKI, T. and NARAHASHI, T. (1959b). Electrical properties of the cockroach giant axon. *J. ins. Physiol.* **3**, 230–42.

YAMASAKI, T. and NARAHASHI, T. (1960). Synaptic transmission in the last abdominal ganglion of the cockroach. *J. ins. Physiol.* **4**, 1–13.

YOUNG, A. C. (1938). The effect of stimulation on the potassium content of *Limulus* leg nerves. *J. Neurophysiol.* **1**, 4–6.

ADDENDA

PAGE 2 A recent study of the structure of the nervous system in *Limulus polyphemus* has revealed that in this xiphosuran the perilemma differs from that of insect species in that the cellular perineurium is absent from peripheral nerve and is only found periodically around the ventral nerve cord (Dumont, Anderson and Chomyn, 1965). The general organization of the connective tissue in the peripheral nerve of this species resembles that found in vertebrate animals. Unlike the condition in insects (Smith and Treherne, 1963) collagen fibres are not confined to the neural lamella surrounding the central nervous system, but are also distributed in the neural lamellae of the peripheral nerves.

PAGE 16 An attempt to define the extent of the extracellular fluid in crab peripheral nerve has been carried out using a method involving the precipitation of iodine within the tissues by silver (Baker, 1965). The resulting electron-dense precipitate was found to be associated with regions containing collagen fibres, in the spaces defined by the mesaxon folds and in space immediately adjacent to the axolemma. Iodine was detected in the mesaxon after an exposure of only 10 sec and was completely washed out within a period of 90 sec. These results serve to emphasize the very rapid nature of the ionic movements which occur in the restricted extracellular spaces of arthropod nervous tissues.

PAGE 32 The intracellular efflux of ^{22}Na from the nerve cords of two insect species (*Periplaneta americana* and *Carausius morosus*) has been shown to be reduced by the presence of ouabain (Treherne, 1966). Uncoupling of the energy supply to the sodium pump, by the addition of cyanide, did not further reduce the sodium efflux from ouabain-treated axons, suggesting that the greater part of the active extrusion of this cation is effected by a single ouabain-sensitive carrier mechanism. The relative insensitivity to ouabain of these insect axons, as compared with squid giant axons (Caldwell and Keynes, 1959), was shown to result from the presence of rather leaky axon membranes in which the carrier-mediated component forms a much smaller part of the total flux. The concentration gradient of sodium across the insect axon membrane can be related to the combined effects of the sodium pump and the passive permeability of the membrane.

PAGE 36   The change in efflux of ^{42}K from the leg nerves of the spider crab, *Maia squinado*, produced by electrical stimulation has been measured using a technique involving as few as 20 shocks (Keynes and Ritchie, 1965). The increased efflux produced by electrical stimulation was found to amount to $2 \cdot 7 \times 10^{-12}$M cm^{-2} impulse^{-1} at 0–$5°$ C and $2 \cdot 1 \times 10^{-12}$M cm^{-2} impulse^{-1} at 22–$25°$ C.

PAGE 83   Stimulation of the intact leg nerve of the spider crab, *Maia squinado*, has been shown to involve an efflux of ninhydrin-positive material which appears to be largely aspartate and glutamate (Baker, 1964). These experiments were carried out in potassium-free media and it is suggested that these amino acids could represent anions which accompany the extrusion of sodium from the axoplasm.

PAGE 108   There has been a recent report of acetylcholine induced contractions in the muscles of the perfused cockroach leg (Kerkut, Shapira and Walker, 1965). This effect was produced with applied solutions of $3 \cdot 4 \times 10^{-6}$M. It is suggested that the negative results obtained by the earlier workers may have resulted from differences in the perfusion techniques employed. The problem as to whether acetylcholine is the natural excitatory transmitter still remains, however, especially in view of the possibility that this substance might act as a non-specific depolarizing agent on the muscles of this preparation.

PAGE 117   Tension measurements and intracellular recording techniques have demonstrated a nervous inhibition of the metathoracic extensor tibiae muscle in the grasshopper (*Romalea microptera*) and the locust (*Schistocerca gregaria*) (Usherwood and Grundfest, 1965). The inhibitory synaptic membranes in these insect preparations exhibit similar electrophysiological and pharmacological properties to those observed in the inhibitory neuro-muscular synapses of the crayfish and lobster. Application of GABA mimics the effect of stimulating the inhibitor axon, while picrotoxin blocks the inhibitory post-synaptic potential and also the effects of GABA.

GABA has also been shown to inhibit glutamate induced contractions in the leg muscles of the cockroach (Kerkut, Shapira and Walker, 1965).

PAGE 117 Application of glutamate ($7 \cdot 0 \times 10^{-7}$M) has been demonstrated to cause contracture in the musculature of cockroach leg preparations (Kerkut, Shapira and Walker, 1965). The possibility that this substance may be a functional transmitter substance at arthropod neuromuscular junctions is strengthened by the observation that glutamate appears in the perfusates of stimulated *Periplaneta* and *Carcinus* muscles (Kerkut *et al.* 1965). The amount of glutamate produced in these experiments was proportional to the number of applied stimuli and was absent from the perfusates of unstimulated muscles.

REFERENCES

BAKER, P. F. (1964). An efflux of ninhydrin-positive material associated with the operation of the Na⁺ pump in intact crab nerve immersed in Na⁺-free solutions. *Biochim. Biophys. Acta*, **88**, 458–60.

BAKER, P. F. (1965). A method for the localization of extracellular space in crab nerve. *J. Physiol.* **180**, 439–47.

CALDWELL, P. C. and KEYNES, R. D. (1959). The effect of ouabain on the efflux of sodium ions from a squid giant axon. *J. Physiol.* **148**, 8P.

DUMONT, J. N., ANDERSON, E. and CHOMYN, E. (1965). The anatomy of the peripheral nerve and its ensheathing artery in the horseshoe crab, *Xiphosura (Limulus) polyphemus. J. Ultrastructure Res.* 13, 38–64.

KERKUT, G. A., LEAKE, L. D., SHAPIRA, A., COWAN, S. and WALKER, R. J. (1965). The presence of glutamate in nerve-muscle perfusates of *Helix*, *Carcinus* and *Periplaneta. Comp. Biochem. Physiol.* **15**, 485–502.

KERKUT, G. A., SHAPIRA, A. and WALKER, R. J. (1965). The effect of acetylcholine, glutamic acid and GABA on contractions of the perfused cockroach leg. *Comp. Biochem. Physiol.* **16**, 37–48.

KEYNES, R. D. and RITCHIE, J. M. (1965). The movements of labelled ions in mammalian non-myelinated nerve fibres. *J. Physiol.* **179**, 333–67.

TREHERNE, J. E. (1966). The effect of ouabain on the efflux of sodium in the nerve cords of two insect species (*Periplaneta americana* and *Carausius morosus*). *J. exp. Biol.* (in the press).

USHERWOOD, P. N. R. and GRUNDFEST, H. (1965). Peripheral inhibition in skeletal muscle of insects. *J. Neurophysiol.* **28**, 497–518.

INDEX

acetylcholine
'bound' and 'free', 94–6
diffusion in nervous tissue, 97–8, 102–3
distribution, 93–8
role in neuromuscular transmission,
108–9, 122
synthesis, 98–100
acetylcholinesterase (cholinesterase)
activity, 100–4
concentration in arthropod nervous
tissues, 100, 107, 122
concentration in skeletal muscles, 108
distribution, 95, 101–2, 107, 111, pl. 2
eserine-insensitive fraction, 102
'pseudo', 101, 107
action potential, 49–53, 69, 120
effects of sodium ions, 44, 55
effects of potassium ions, 53–5
other inorganic ions, 55–62
of muscle fibres, 62
adrenalin, 110–11
Aeschna, neuromuscular junction, 11, 12
after-potential
effects of ions, 62–5
effects of extracellular system, 121
alanine
concentration in blood, 82
concentration in nervous tissues, 15,
81, 82, 83, 84
metabolism, 83, 84, 85–6
β-alanine, as transmitter substance, 117
amino acids
concentration in nervous tissues, 80–
3, 122
concentration in blood, 80, 82, 122
metabolism, 83–6
role in osmotic balance of axoplasm,
13, 15
as transmitter substances, 114–18
γ-aminobutyric acid (GABA)
concentration in nervous tissues, 81,
83, 84, 85
molecular basis of activity, 111, 116
synthesis, 86
as transmitter substance, 86
Amphipods, trehalose, 73

'anaerobic resting potential', 70
anions
effects on nervous activity, 45–6
extracellular fraction in nervous
tissues, 23, 26
fixed anion groups, 25–6, 27, 37, 38, 40
Annelids, 5-hydroxytryptamine, 112
Anthopera (solitary bee), blood trehalose,
73
Apis mellifica, see bee
arachnids
acetylcholine, 95
nervous systems, 1, 2
trehalose, 73
arginine, in central nervous tissues of
bee, 81
Armadillidium vulgare (Isopod), structure
of nerves, 3, 10
arsenite, inhibition of enzymes by, 84–5
Artemia saline (brine shrimp), sodium in
blood of, 21
aspartate
concentration in blood, 82
metabolism, 83, 84
in nervous tissues, 81, 82, 83
role in osmotic balance, 13, 15
Astacus fluviatilis, acetylcholine content
of nervous tissues, 95
ATP (adenosine triphosphate)
combination with calcium, 39
and sodium extrusion, 30, 70–1, 121
in synthesis of acetylcholine, 99
ATP-ase, in nerve, 71
atropine, and cardiac-accelerator fibres,
109
axolemma, relation with Schwann cells,
63, 71
axons
glial cell sheaths, 4, 6, 63
structure, 4, 7, 8, 10
structure of surface, 90–1
axoplasm, 8, 10, 11
extracellular fluid and, 39
effects of stimulation on osmotic concentration, 18
viscosity, 8

mitochondria (*cont.*)
glial cells, 5
perineurium, 3, 78, pl. 1
molluscs, 5-hydroxytryptamine content,
112
mosquito larvae, 28
glycogen in nervous tissues, 77
motor end-plates, acetylcholinesterase
activity, 108
motor fibres, *see* nerve fibres
moulting cycle, in Crustacea, blood
calcium changes, 57
mouse brain, glutamic decarboxylase,
86, 115
mucopolysaccharides, in nerve, 27, 38,
79; in neural lamella, 3, 78–9
Musca domestica (housefly)
acetylcholine in, 95, 98–9
acetylcholinesterase in, 100–1
muscle fibres, action potential, 62;
resting potential, 48
myelination, in Crustacea and vertebrates, 7

nerve cell body, 7, 8
γ-aminobutyric acid, 116
nerve cords, structure of, 1
nerve fibres, 7, 9, 13, 18, 33
inhibitory, 94, 117, 118, 119
motor, 94, 108, 118
sensory, 94, 114, 118
nerve sheath, 2–3, 13, pl. 1 (*a*) and 3 (*a*)
acetylcholine, 98, 105
acetylcholinesterase, 104
effect of removal, on potassium depolarization, 42, 43, 44; on exchange of ions, 25, 26–7, 44; on
conduction in sodium-deficient
solution, 45
see also neural lamella, perineurium
neural lamella, 2–3, 4, 78–9, pl. 1
permeability, 77, 78
neurofibrillae, 8
neurofilaments, 10
neuromuscular junction, 11–12, pl. 2 (*b*)
acetylcholinesterase activity, in insects, 101
transmission, effects of catecholamines, 111; effects of acetylcholine,
108–9, 122; effects of γ-aminobutyric acid, 117, 118, 5-hydroxy-
tryptamine, 113, 114, Substance I,
118
neurones, 7–10
of stretch receptor, *see* stretch receptor

neuropile, 4, 8, acetylcholinesterase in,
101, 107, pl. 3
nor-adrenalin, 110
nucleic acids, 39

oil droplets in axoplasm, 8
opacity of nerve, 18
Orconectes virilis
5-hydroxytryptamine in nerve cord,
113
resting and action potentials, 47, 50,
57
see also crayfish
ouabain, 66, 71
α-oxoglutaric dehydrogenase, 84–5
oxygen
consumption, 66
effect of lack, on onset of fatigue, 69;
on hydrolysis of glycogen, 76; on
potassium loss, 35

peptides, in brain of bee, 81–2
pericardial organ of Crustacea, 112
perikaryon, 4, 8, 10, 101, 102
perineurium, 2–3, 4, pl. 1
acetylcholinesterase in, 101, 104
glycogen in, 76, 77, 78
lipids in, 88
Periplaneta americana (cockroach)
blood amino acids, 82; inorganic ions,
22, 27
nerve action potential, 51; after
potential, 62–5; amino acids in,
81, 82, 83; barium, 62; calcium,
37, 58–9; carbohydrates, 73–4, 76,
77; chloride, 40; inorganic ions,
22, 27; lipids, 88; membrane
capacity, 52; potassium, 21, 32–4,
43, 44, 47, 83; sodium, 24–32, 43,
45, 49, 50, 52, 53, 54, 55–6; water
relations, 14–15, 16
structure of nervous tissues, 2, 3, 4,
7, 8, 9, 10, 16, pls. 1, 2, 3 (*a*)
transmitter substances, acetylcholine,
95, 98, 99, 100, 101, 102–3, 104,
107; catecholamines, 110–11, 112;
γ-aminobutyric acid, 115
permeability of membranes
changes, 59–60
effect of γ-aminobutyric acid, 116,
119
energy requirement, 66
permeability constant, 28
Petrobius maritimus (Apterygota), blood,
20

Phasmida, inorganic ions in blood, 21
phosphate
in blood, 22
high energy, 70–1, 121; *see also* ATP
in nerve, 15, 22
phosphoenolpyruvate, 85
phospholipases, action on axon surface, 90, 91
phospholipid-protein structure of axon membrane, 60, 90–2
phospholipids
combination with calcium, 39
in nerves of different species, 89
structural integrity and nervous activity, 90
phosphorylase activity, 76
phosphorylation (oxidative), and sodium extrusion, 30, 70–1
picrotoxin, 118
potassium
active uptake, 34
and after potential, 63, 65
balanced by dicarboxylic amino acids, 83
concentration in blood, 20, 21, 22, 27; of *Carausius*, 28, 30, 83
concentration in nervous tissues, 15, 21, 22, 23; extra- and intracellular fractions, 23, 27, 28, 30, 35, 36, 47
depolarization of nerve by, *see* depolarization
effect of concentration on free acetylcholine content, 96; on injury potential, 41, 53; on sodium extrusion, 30, 31, 32; on resting and action potentials, 53–5
equilibrium potential, 47; and resting potential of axons, 46, 47, 48, 55, 120
exchanges of, 27, 28, 32–7
interactions with calcium, 41–2, 59
heat of ionic exchange with sodium, 67–8
intracellular concentration and resting potential, 48, 120
movements during electrical activity, 18, 36–7, 51, 52, 67
permeability of post-synaptic membrane, 116
uptake by linked-ion pump, 30, 32, 34, 36, 46, 70–1, 121
prawns, structure of nerve fibres, 7
Procambarus alleni, resting and action potentials, 47, 50

proline, concentration in blood, 82; in nervous tissues, 81, 82, 84
proteases, action on axon surface, 90–1
protein metabolism, 86–7
protein-phospholipid structure of axon membrane, 60, 90–2
pyruvate, 85–6
pyruvic decarboxylase, 84, 86

rabbit
choline acetylase activity in brain, 99
lipids in, 89
rat, amino acids in brain, 81
recovery processes, after stimulation of nerve, 68, 69
repolarization, 60, 116
resting potential, 46–9, 50
anaerobic and aerobic fractions, 70
effects of external potassium concentration, 53–5
effects of other inorganic ions, 56–62
of muscle fibres, 48
relation to potassium equilibrium potential, 47–8, 120
Rhodnius prolixus
acetylcholinesterase, 101
lipids in nervous tissue, 88
structure of nervous tissues, 2, 3, 5, 7
ribonucleic acid, in stretch receptor, 87, 88
ribosomes, 8
Romalea microptera
inorganic ion content of blood and nerve, 21, 22
water content of nerve, 14
rubidium, effects on nervous activity, 45

Schwann cells (lemnoblasts), 6, 7, 11, 71
relation with axolemma, 63
sensory fibres, *see* nerve fibres
Sepia, ion exchanges in giant axons, 32
serine, in central nerve tissue of bee, 81
sodium
active uptake, by nerve cord of *Carausius*, 27–30, 40, 120
active extrusion, 30, 32, 34, 36, 46, 69, 70–1
concentration in blood of arthropod species, 20, 21, 22, 27; of *Carausius*, 24, 28, 30, 48, 49
concentration in nervous tissue, 15, 21, 22; extra- and intracellular fractions, 23, 25, 27, 28, 30, 36, 50
effects of external concentration on excitability of nerve, 44–5; on